Anywhere-Anytime Signals and Systems Laboratory

From MATLAB to Smartphones

Third Edition

Synthesis Lectures on Signal Processing

Editor
José Moura, *Carnegie Mellon University*

Synthesis Lectures in Signal Processing publishes 80- to 150-page books on topics of interest to signal processing engineers and researchers. The Lectures exploit in detail a focused topic. They can be at different levels of exposition-from a basic introductory tutorial to an advanced monograph-depending on the subject and the goals of the author. Over time, the Lectures will provide a comprehensive treatment of signal processing. Because of its format, the Lectures will also provide current coverage of signal processing, and existing Lectures will be updated by authors when justified.

Lectures in Signal Processing are open to all relevant areas in signal processing. They will cover theory and theoretical methods, algorithms, performance analysis, and applications. Some Lectures will provide a new look at a well established area or problem, while others will venture into a brand new topic in signal processing. By careful reviewing the manuscripts we will strive for quality both in the Lectures' contents and exposition.

Anywhere-Anytime Signals and Systems Laboratory: From MATLAB to Smartphones, Third Edition
Nasser Kehtarnavaz, Fatemeh Saki, Adrian Duran, and Arian Azarang
2020

Reconstructive-Free Compressive Vision for Surveillance Applications
Henry Braun, Pavan Turaga, Andreas Spanias, Sameeksha Katoch, Suren Jayasuriya, and Cihan Tepedelenlioglu
2019

Smartphone-Based Real-Time Digital Signal Processing, Second Edition
Nasser Kehtarnavaz, Abhishek Sehgal, Shane Parris
2018

Anywhere-Anytime Signals and Systems Laboratory: from MATLAB to Smartphones, Second Edition
Nasser Kehtarnavaz, Fatemeh Saki, and Adrian Duran
2018

Nonlinear Source Separation
Luis B. Almeida
2006

Spectral Analysis of Signals: The Missing Data Case
Yanwei Wang, Jian Li, and Petre Stoica
2006

Anywhere-Anytime Signals and Systems Laboratory: From MATLAB to Smartphones, Third Edition

Nasser Kehtarnavaz, Fatemeh Saki, Adrian Duran, and Arian Azarang

ISBN: 978-3-031-01414-7   paperback
ISBN: 978-3-031-02542-6   ebook
ISBN: 978-3-031-00335-6   hardcover

DOI 10.1007/978-3-031-02542-6

A Publication in the Springer series
SYNTHESIS LECTURES ON SIGNAL PROCESSING

Lecture #18
Series Editor: José Moura, *Carnegie Mellon University*
Series ISSN
Print 1932-1236   Electronic 1932-1694

Anywhere-Anytime Signals and Systems Laboratory

From MATLAB to Smartphones

Third Edition

Nasser Kehtarnavaz, Fatemeh Saki, Adrian Duran, and Arian Azarang
University of Texas at Dallas

SYNTHESIS LECTURES ON SIGNAL PROCESSING #18

ABSTRACT

A typical undergraduate electrical engineering curriculum incorporates a signals and systems course. The widely used approach for the laboratory component of such courses involves the utilization of MATLAB to implement signals and systems concepts. This book presents a newly developed laboratory paradigm where MATLAB codes are made to run on smartphones which are possessed by nearly all students. As a result, this laboratory paradigm provides an anywhere-anytime hardware platform or processing board for students to learn implementation aspects of signals and systems concepts. The book covers the laboratory experiments that are normally covered in signals and systems courses and discusses how to run MATLAB codes for these experiments as apps on both Android and iOS smartphones, thus enabling a truly mobile laboratory paradigm. A zipped file of the codes discussed in the book can be acquired via the website http://sites.fastspring.com/bookcodes/product/SignalsSystemsBookcodesThirdEdition.

KEYWORDS

smartphone-based signals and systems laboratory, anywhere-anytime platform for signals and systems courses, from MATLAB to smartphones

Contents

Preface

A typical undergraduate electrical engineering curriculum incorporates a signals and systems course where students normally first encounter signal processing concepts of convolution, Fourier series, Fourier transform, and discrete Fourier transform. For the laboratory component of such courses, the conventional approach has involved a laboratory environment consisting of computers running MATLAB codes. There exist several lab textbooks or manuals for the laboratory component of signals and systems courses based on MATLAB, e.g., *An Interactive Approach to Signals and Systems Laboratory* by Kehtarnavaz, Loizou, and Rahman; *Signals and Systems Laboratory with MATLAB* by Palamides and Veloni; *Signals and Systems: A Primer with MATLAB* by Sadiku and Ali; and *Signals and Systems* by Mitra.

The motivation for writing this lab textbook/manual has been to provide an alternative laboratory paradigm to the above conventional laboratory paradigm by using smartphones as a truly mobile anywhere-anytime hardware platform or processing board for students to run signals and systems codes written in MATLAB on them. This approach or laboratory paradigm eases the requirement of using a dedicated laboratory room for signals and systems courses and allows students to use their own computers/laptops and smartphones/tablets as the hardware platform to learn the implementation aspects of signals and systems concepts. It is worth stating that this book is only meant as an accompanying lab book to signals and systems textbooks and is not meant to be used as a substitute for these textbooks.

The challenge in developing this alternative approach has been to limit the programming language required from students to MATLAB and not requiring them to know any other programming language. MATLAB is extensively used in engineering departments and students are often expected to use it for various courses they take during their undergraduate studies.

The above challenge is met here by using the smartphone software tools that are publicly available. The software development environments of smartphones (both Android and iOS) are free of charge and students can download and place them on their own computers/laptops. In this lecture series book, we have developed the software shells that allow students to run MATLAB codes on their own smartphones/tablets as apps. In the first edition of the book, the implementation was done on Android smartphones. In the second edition, in addition to Android smartphones, the implementation was done on iOS smartphones. Due to various updates that have taken place in MATLAB and in smartphone software tools, this third edition is written to address incompatibility errors caused by the older versions of the software tools when running the codes in the previous editions.

The book chapters correspond to the following labs for a semester-long lab course: (1) introduction to MATLAB programming; (2) smartphone development tools (both Andorid and iOS); (3) use of MATLAB Coder to generate C codes from MATLAB and running C codes on smartphones; (4) linear time-invariant systems and convolution; (5) Fourier series; (6) continuous-time Fourier transform; and (7) digital signals and discrete Fourier transform. A typical signals and systems laboratory course or component covers the labs associated with subjects (4)–(7).

Finally, it is to be noted that the codes discussed in the book can be acquired from this third-party website http://sites.fastspring.com/bookcodes/product/ SignalsSystemsBookcodesThirdEdition.

Nasser Kehtarnavaz, Fatemeh Saki, Adrian Duran, and Arian Azarang
Summer 2020

CHAPTER 1

Introduction to MATLAB

MATLAB is a programming environment that is widely used to solve engineering problems. There are many online references on MATLAB that one can read to become familiar with this programming environment. This chapter is only meant to provide an overview or a brief introduction to MATLAB. Screenshots are used to show the steps to be taken and configuration options to set when using the Windows operating system.

1.1   STARTING MATLAB

Assuming MATLAB is installed on the laptop or computer used, select MATLAB from the Start bar of Windows, as illustrated in Figure 1.1. After starting MATLAB, a window called MATLAB desktop appears, see Figure 1.2, which contains other sub-windows or panels. The panel **Command Window** allows interactive computation to be conducted. Suppose it is desired to compute $3 + 4 * 6$. This is done by typing it at the prompt command denoted by `>>` ; see Figure 1.3. Since no output variable is specified for the result of $3 + 4 * 6$, MATLAB returns the value in the variable `ans` , which is created by MATLAB. Note that `ans` is always overwritten by MATLAB, so if the result is used for another operation, it needs to be assigned to a variable, for example x $= 3 + 4 * 6$.

In practice, a sequence of operations is usually performed to achieve a desired output. Often, a so-called m-file script is used for this purpose. An m-file script is a simple text file where MATLAB commands are listed. Figure 1.4 shows how to start a new script. In the **HOME** menu, locate the *New Script* tab under *New → Script*, or *Ctrl+N* to create a blank script under the panel **EDITOR**. When a new script is opened, it looks as shown in Figure 1.5. A script can be saved using a specified name in a desired location. An m-file script is saved with '.m' extension. When such a file is run, MATLAB reads the commands and executes them as though there were the MATLAB commands and operations. The following section provides more details on the MATLAB commands and operations.

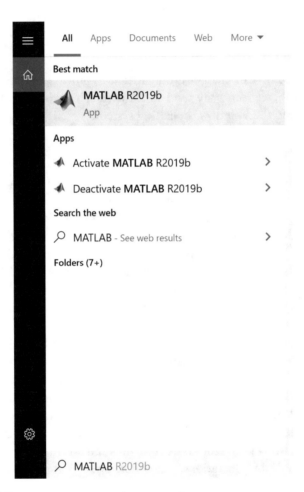

Figure 1.1: MATLAB appearance in windows start bar.

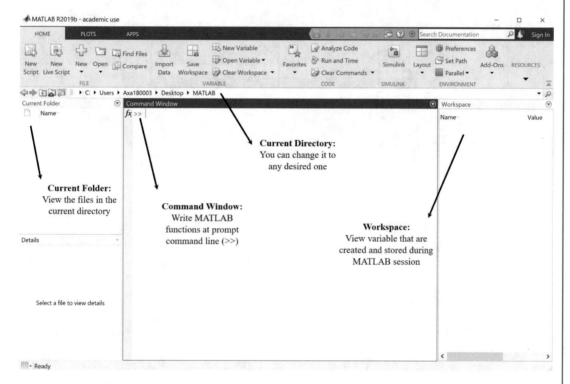

Figure 1.2: **MATLAB** interface window.

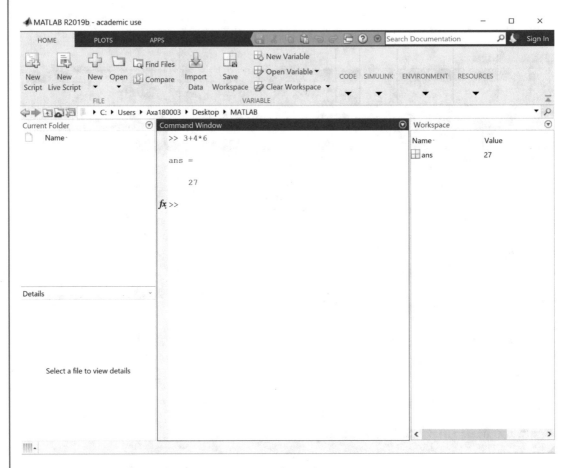

Figure 1.3: A simple computation in command window.

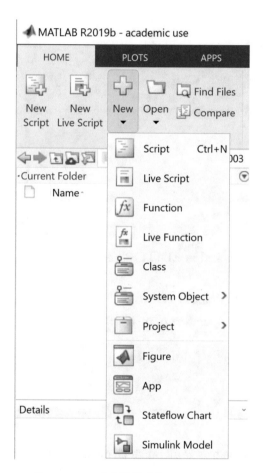

Figure 1.4: Starting a new m-file script in MATLAB.

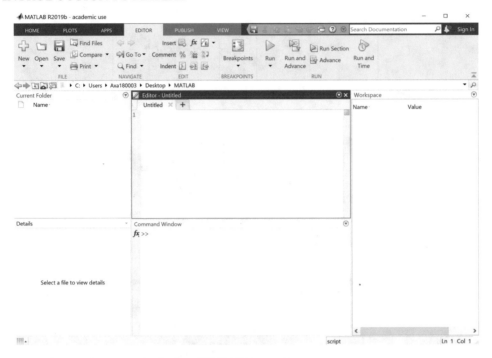

Figure 1.5: **An m-file script docked in EDITOR panel.**

1.1.1   ARITHMETIC OPERATIONS

There are four basic arithmetic operators in m-files:

```
+ addition
- subtraction
* multiplication
/ division (for matrices, it also means inversion)
```

The following three operators work on an element-by-element basis:

```
.* multiplication of two vectors, element-wise
./ division of two vectors, element-wise
.^ raising all the elements of a vector to a power
```

As an example, to evaluate the expression $a^3 + \sqrt{bd} - 4c$, where $a = 1.2$, $b = 2.3$, $c = 4.5$, and $d = 4$, type the following commands in the **Command Window** to get the answer (`ans`):

```
>> a=1.2;
>> b=2.3;
>> c=4.5;
>> d=4;
>> a^3+sqrt(b*d)-4*c
ans =
 -13.2388
```

Note the semicolon after each variable assignment. If the semicolon is omitted, the interpreter echoes back the variable value.

1.1.2   VECTOR OPERATIONS

Consider the vectors $x = [x_1, x_2, \ldots, x_n]$ and $y = [y_1, y_2, \ldots, y_n]$. The following operations indicate the resulting vectors:

$$x. * y = [x_1 y_1, x_2 y_2, \ldots, x_n y_n]$$
$$x./y = \left[\frac{x_1}{y_1}, \frac{x_2}{y_2}, \ldots, \frac{x_n}{y_n} \right]$$
$$x. \wedge p = \left[x_1^p, x_2^p, \ldots, x_n^p \right].$$

Considering that the boldfacing of vectors/matrices are not used in .m files, in the notation adopted in this book, no boldfacing of vectors/matrices is shown to retain notation consistency with .m files.

The arithmetic operators $+$ and $-$ can be used to add or subtract matrices, vectors, or scalars. Vectors denote 1-dimensional (1-D) arrays and matrices denote multi-dimensional arrays. For example:

```
>> x=[1,3,4]
>> y=[4,5,6]
>> x+y
ans=
 5 8 10
```

In this example, the operator $+$ adds the elements of the vectors x and y, element by element, assuming that the two vectors have the same dimension, in this case 1×3 or one row with three columns. An error occurs if one attempts to add vectors having different dimensions. The same applies for matrices.

To compute the dot product of two vectors (in other words, $\sum_i x_i y_i$), use the multiplication operator `*` as follows:

```
>> x*y'
ans =
  43
```

Note the single quote after `y` denotes the transpose of a vector or a matrix.

An element-by-element multiplication of two vectors (or two arrays) is computed by the following operator:

```
>> x .* y
ans =
   4 15 24
```

That is, `x .* y` means $[1 \times 4, \ 3 \times 5, \ 4 \times 6] = [4 \ 15 \ 24]$.

1.1.3 COMPLEX NUMBERS

MATLAB supports complex numbers. The imaginary number is denoted with the symbol *i* or *j*, assuming that these symbols have not been used any other place in the program. It is critical to avoid such symbol conflicts for obtaining correct outcomes. Enter the following and observe the outcomes:

```
>> z=3 + 4i % note the multiplication sign `*´ is not needed after 4
>> conj(z)  % computes the conjugate of z
>> angle(z) % computes the phase of z
>> real(z)  % computes the real part of z
>> imag(z)  % computes the imaginary part of z
>> abs(z)   % computes the magnitude of z
```

One can also define an imaginary number with any other user-specified variables. For example, in the following manner:

```
>> img=sqrt(-1)
>> z=3+4*img
>> exp(pi*img)
```

1.1.4 ARRAY INDEXING

In m-files, all arrays (vectors) are indexed starting from 1; in other words, x(1) denotes the first element of the array x. Note that arrays are indexed using parentheses `(.)` and not square brackets `[.]`, as done in C/C++. To create an array featuring the integers 1–6 as elements, enter:

```
>> x=[1,2,3,4,5,6]
```

Alternatively, the notation `` `:` `` can be used as follows:

```
>> x=1:6
```

This notation creates a vector starting from 1–6, in steps of 1. If a vector from 1–6 in steps of 2 is desired, then type:

```
>> x=1:2:6
ans =
  1 3 5
```

Additional examples are listed below:

```
>> ii=2:4:17
>> jj=20:-2:0
>> ii=2:(1/10):4
```

One can easily extract numbers in a vector. To concatenate an array, the example below shows how to use the operator `` `[ ]` ``:

```
>> x=[1:3 4 6 100:110]
```

To access a subset of this array, type the following:

```
>> x(3:7)
>> length(x)    % gives the size of the array or vector
>> x(2:2:length(x))
```

1.1.5   ALLOCATING MEMORY

Memory can get allocated for 1-D arrays (vectors) using the command or function `zeros`. The following command allocates memory for a 100-dimensional array:

```
>> y=zeros(100,1);
>> y(30)
ans =
   0
```

The function `zeros(n,m)` creates an n×m matrix with all 0 elements. One can allocate memory for 2-dimensional (2-D) arrays (matrices) in a similar fashion. The command or function

```
>> y=zeros(4,5)
```

defines a 4 by 5 matrix.

Similar to the command `zeros`, the command `ones` can be used to define a vector containing all ones. For example,

```
>> y=ones(1,5)
ans=
   1 1 1 1 1
```

1.1.6   SPECIAL CHARACTERS AND FUNCTIONS

Here is an example of the function `length`:

```
>> x=1:10;
>> length(x)
ans =
   10
```

The function `find` returns the indices of a vector that are non-zero. For example, `I = find(x>4)` finds all the indices of x greater than 4. Thus, for the above example:

```
>> find(x> 4)
ans =
   5 6 7 8 9 10
```

1.1.7   CONTROL FLOW

m-files have the following control flow constructs:

Table 1.1: Some widely used special characters used in m-files

Symbol	Meaning
pi	$\pi(3.14.....)$
^	Indicates power (for example, $3^2 = 9$)
NaN	Not-a-number, obtained when encountering undefined operations, such as 0/0
Inf	Represents $+\infty$
;	Indicates the end of a row in a matrix; also used to suppress printing on the screen (echo off)
%	Comments—anything to the right of % is ignored by the .m file interpreter and is considered to be comments
'	Denotes transpose of a vector or a matrix; also used to define string, for example, $str1$ = 'DSP'
...	Denotes continuation; three or more periods at the end of a line continue current function to next line

Table 1.2: Some widely used functions

Function	Meaning
sqrt	Indicates square root, for example, $sqrt(4) = 2$
abs	Absolute value |.|, for example, $abs(-3) = 3$
length	$length(x)$ gives the dimension of the array x
sum	Finds sum of the elements of a vector
find	Finds indices of nonzero

- `if` statements
- `switch` statements
- `for` loops
- `while` loops
- `break` statements

The constructs `if`, `for`, `switch`, and `while` need to terminate with an `end` statement. Examples are provided below:

if

```
>> x=-3;
if x>0
 str='positive'
elseif x<0
 str='negative'
elseif x== 0
 str='zero'
else
 str='error'
end
```

See the value of `str` after running the above code.

while

```
>> x=-10;
while x<0
 x=x+1;
end
```

See the value of x after running the above code.

for loop

```
>> x=0;
for j=1:10
 x=x+j;
end
```

The above code computes the sum of all the numbers from 1–10.

break

With the break statement, one can exit early from a `for` or a `while` loop. For example:

```
>> x=-10;
while x<0
 x=x+2;
 if x = = -2
```

Table 1.3: Relational operators

Symbol	Meaning
<=	Less than equal
<	Less than
>=	Greater than equal
>	Greater than
==	Equal
~=	Not equal

Table 1.4: Logical operators

Symbol	Meaning
&	AND
|	OR
~	NOT

```
    break;
  end
end
```

Some of the supported relational and logical operators are listed below.

1.1.8   PROGRAMMING IN MATLAB

Open a new script file as displayed in Figures 1.3 and 1.4. Save it first in a desired directory. Then, write your MATLAB code and press **Run** button from the **EDITOR** panel. For instance, to write a program to compute the average (mean) of a vector x, the program should use the vector x as its input and return the average value. To write this program, follow the steps outlined below.

Type the following in an empty window:

```
x=1:10
L=length(x);
sum=0;
for j=1:L
sum=sum+x(j);
```

Figure 1.6: m-file script interactive window after running the program average.

```
end
y=sum/L % y returns the average of x
```

From the **EDITOR** panel, go to *save* → *Save As* and enter **average.m** for the filename. Then, click on the **Run** button to run the program. Figure 1.6 shows the MATLAB interactive window after running the program.

1.1.9   SOUND GENERATION

Assuming the computer used has a sound card, one can use the function `sound` to play back speech or audio files through its speakers. That is, the function `sound(y,FS)` sends the signal in a vector `y` (with sample frequency FS) out to the speaker. Stereo sounds are played on platforms that support them, with y being an N-by-2 matrix.

Type the following code and listen to a 400 Hz tone:

```
>> t=0:1/8000:1;
>> x=cos(2*pi*400*t);
>> sound(x,8000);
```

Now generate a noise signal by typing:

```
>> noise=randn(1,8000);  %  generate 8000 samples of noise
>> sound(noise,8000);
```

The function `randn` generates Gaussian noise with zero mean and unit variance.

1.1.10   LOADING AND SAVING DATA

One can load or store data using the commands `load` and `save`. To save the vector `x` of the above code in the file **data.mat**, type:

```
>> save('data.mat', 'x')
```

To retrieve the data previously saved, type:

```
>> load data
```

The vector `x` gets loaded in memory. To see memory contents, use the command `whos`:

```
>> whos
Variable  Dimension   Type
x         1x8000      double array
```

The command `whos` gives a list of all the variables currently in memory, along with their dimensions and data type. In the above example, `x` contains 8000 samples.

To clear up memory after loading a file, type `clear all` when done. This is important because if one does not clear all the variables, conflicts can occur with other codes using the same variables.

1.1.11   READING WAVE AND IMAGE FILES

In MATLAB, one can read data from different file types (such as .wav, .jpeg, and .bmp) and load them in a vector.

To read an audio data file with .wav extension, use the following command:

```
>> [y,Fs]=audioread('filename')
```

This command reads a .wav file specified by the string `filename` and returns the sampled data in `y` with the sampling rate of Fs (in Hz).

To read an image file, use the following command:

```
>> [y]=imread('filename')
```

This command reads a grayscale or color image from the string `filename` and returns the image data in the array `y` .

1.1.12   SIGNAL DISPLAY

Graphical tools are available in MATLAB to display data in a graphical form. Throughout the book, signals in both the time and frequency domains are displayed using the function plot,

```
>> plot(x,y)
```

This function creates a 2-D line plot of the data in `y` vs. corresponding `x` values.

1.2   MATLAB PROGRAMMING EXAMPLES

In this section, several simple MATLAB programs are covered.

1.2.1   SIGNAL GENERATION

In this example, let us see how to generate and display continuous-time signals in the time domain. One can represent such signals as a function of time. For simulation purposes, a representation of time t is needed. Note that the time scale is continuous while computers handle operations in a discrete manner. Continuous-time simulation is achieved by considering a very small time interval. For example, if a 1-s duration signal in millisecond (ms) increments (time interval of 0.001 s) is considered, then one sample every 1 ms or a total of 1000 samples are generated for the entire signal leading to a continuous signal simulation. This continuous-time signal approximation or simulation is used in later chapters. It is important to note that a finite number of samples is involved in the simulation of a continuous-time signal, and thus to differentiate a continuous-time signal from a discrete-time signal, a much higher number of samples per second for a continuous-time signal needs to be used (very small time interval).

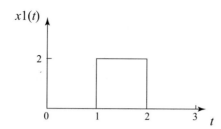 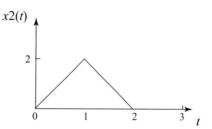

Figure 1.7: Continuous-time signal.

Figure 1.7 shows two continuous-time signals $x1(t)$ and $x2(t)$ with a duration of 3 s. By setting the time interval dt to 0.001 s, there is a total of 3000 samples at $t = 0, 0.001, 0.002, 0.003, \ldots, 2.999$ s.

Note that throughout the book, the notations dt, delta, and Δ are used interchangeably to denote the time interval between samples.

The signal $x1(t)$ can be represented mathematically as follows:

$$x1(t) = \begin{cases} 0 & 0 \leq t < 1 \\ 1 & 1 \leq t < 2 \\ 0 & 2 \leq t < 3. \end{cases} \tag{1.1}$$

To simulate this analog or continuous-time signal, use the MATLAB functions `ones` and `zeros` . The signal value is zero during the first second, which means the first 1000 samples are zero. This portion of the signal is simulated with the function `zeros(1,1000)` . In the next second (next 1000 samples), the signal value is 2, and this portion is simulated by the function `2*ones(1,1000)` . Finally, the third portion of the signal is simulated by the function `zeros(1,1000)` . In other words, the entire duration of the signal is simulated by the following .m file function:

```
x1=[ zeros(1,1/dt) 2*ones(1,1/dt) zeros(1,1/dt)]
```

The signal $x2(t)$ can be represented mathematically as follows:

$$x2(t) = \begin{cases} 2t & 0 \leq t < 1 \\ -2t + 4 & 1 \leq t < 2 \\ 0 & 2 \leq t < 3. \end{cases} \tag{1.2}$$

A linearly increasing or decreasing vector can thus be used to represent the linear portions. The time vectors for the three portions or segments of the signal are `0:dt:1-dt`, `1:dt:2-dt` , and `2:dt:3-dt` . The first segment is a linear function corresponding to a time vector with a slope of 2; the second segment is a linear function corresponding to a time vector with a slope of -2

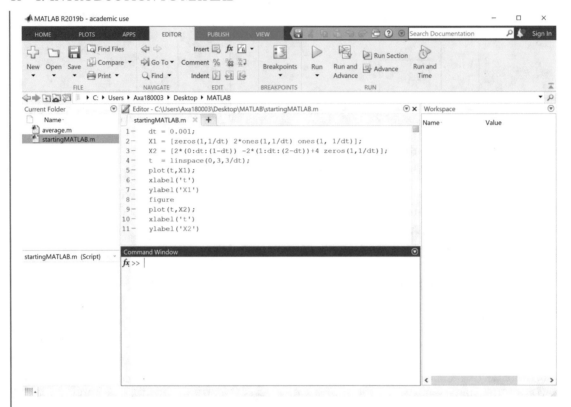

Figure 1.8: **MATLAB** code of signal generation example.

and an offset of 4; and the third segment is simply a constant vector of zeros. In other words, the entire duration of the signal for any value of *dt* can be simulated by the following .m file function:

```
x2=[2*(0:dt:(1-dt)) -2*(1:dt:(2-dt))+4 zeros(1,1/dt)]
```

Figures 1.8 and 1.9 show the MATLAB code and the plot of the above signal generation, respectively. Signals can be displayed using the function `plot(t,data)`. For proper plotting, first the correct `t` vector needs to be generated. Here this is done by using the function `linspace`:

```
>> t=linspace(0,E,N)
```

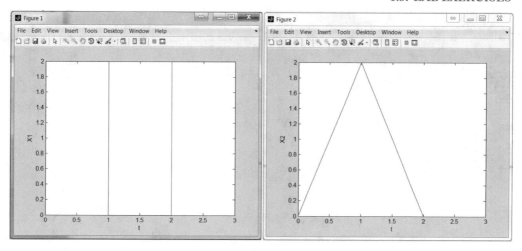

Figure 1.9: Signal plots.

This function generates a vector `t` of `N` points linearly spaced between and including `0` and `E`, where `N` is equal to `E/dt`.

1.2.2   GENERATING A PERIODIC SIGNAL

In this example, a simple periodic signal is generated. This example involves generating a periodic signal in textual mode and displaying it graphically. The shape of the signal (`sin`, `square`, `triangle`, or `sawtooth`) can be modified as well as its frequency and amplitude by using appropriate control parameters. The MATLAB code and the plots generated by it are shown in Figures 1.10 and 1.11, respectively.

Now consider an m-file code to generate four types of waveforms using the functions `sin`, `square`, and `sawtooth`. To change the amplitude and frequency of the waveforms, two control parameters named Amplitude (`A`) and Frequency (`f`) are used. Waveform Type (`w`) is another parameter used for controlling the waveform type. With this control parameter, one can select from multiple inputs. Finally, the waveforms are displayed by using the function `plot`.

1.3   LAB EXERCISES

1. Write an m-file code to add all the numbers corresponding to the even indices of an array. For instance, if the array x is specified as x = [1, 3, 5, 10], then 13 (= 3 + 10) should be returned. Use the program to find the sum of all even integers from 1–1000. Run your code. Also, rewrite the code where x is the input vector and y is the sum of all the numbers corresponding to the even indices of x.

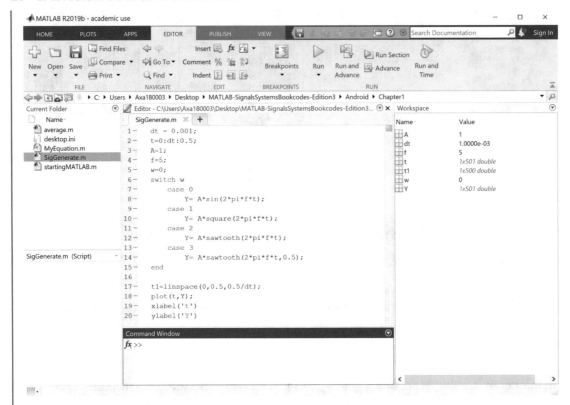

Figure 1.10: Periodic signal generation example.

2. Explain the operation performed by the following .m file:

```
L=length(x);
for j=1:L
if x(j) < 0
x(j)=-x(j);
end
end
```

Rewrite this program without using a `for` loop.

3. Write a .m file code that implements the following hard-limiting function:

$$x(t) = \begin{cases} 0.2 & t \geq 0.2 \\ -0.2 & t < 0.2. \end{cases} \qquad (1.3)$$

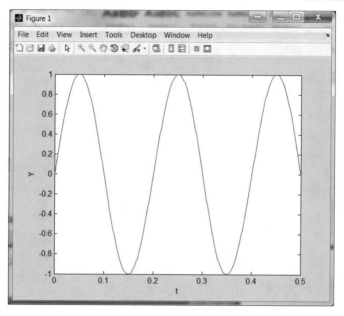

Figure 1.11: Plot of periodic sinusoid signal.

For t, use 1000 random numbers generated by using the function `rand`.

4. Write a MATLAB code to generate two sinusoid signals with the frequencies f1 Hz and f2 Hz and the amplitudes A1 and A2, based on a sampling frequency of 8000 Hz with the number of samples being 256. Set the frequency ranges from 100–400 Hz and set the amplitude ranges from 20–200. Generate a third signal with the frequency `f3 = (mod (lcm (f1, f2), 400) + 100)` Hz, where `mod` and `lcm` denote the modulus and least common multiple operation, respectively, and the amplitude A3 is the sum of the amplitudes A1 and A2. Use the same sampling frequency and number of samples as specified for the first two signals. Display all the signals using the legend on the same waveform graph and label them accordingly.

CHAPTER 2

Software Development Tools

This chapter covers the steps that need to be taken in order to install the software tools for running C codes on Android and iPhone smartphones. Later in Chapter 3, it will be described how to convert MATLAB codes to C codes.

The Android development environment used here is the IntelliJ IDEA-based Android Studio Bundle (called Android Studio). C codes are made available to the Android Java environment through the use of the Java Native Interface (JNI) wrapper. Thus, it is also necessary to install the Android Native Development Kit (NDK). This development kit allows one to write C codes, compile, and debug them on an emulated Android platform or on an actual Android smartphone/tablet.

Screenshots are used to show the steps and configuration options involved in the installation when using the Windows operating system. The same software tools are also available for other operating systems.

2.1 ANDROID TOOLS INSTALLATION STEPS

This section covers the installation of the necessary software packages for Android app development. Start by creating a directory where the tools are to be installed. A generic directory of *C:\Android* is used here and the setup is done such that all Android development related files are placed within the directory *C:\Android*. Before running Android Studio, the latest Java JDK needs to be installed.

2.1.1 JAVA JDK

If the Java Development Kit (JDK) is not already installed on your computer or you do not have the latest version, download it from Oracle's website and follow the installation steps indicated by the installer. The latest JDK package at the time of this writing can be found on the Oracle's website at http://www.oracle.com/technetwork/java/javase/downloads/index.html.

Click on the JDK *Download* button in the section *Oracle JDK* as shown in Figure 2.1 and you will be directed to the page shown in Figure 2.2. From the list of supported platforms, select the correct version for your operating system. For example, if you are running a 64-bit operating system, select the appropriate package.

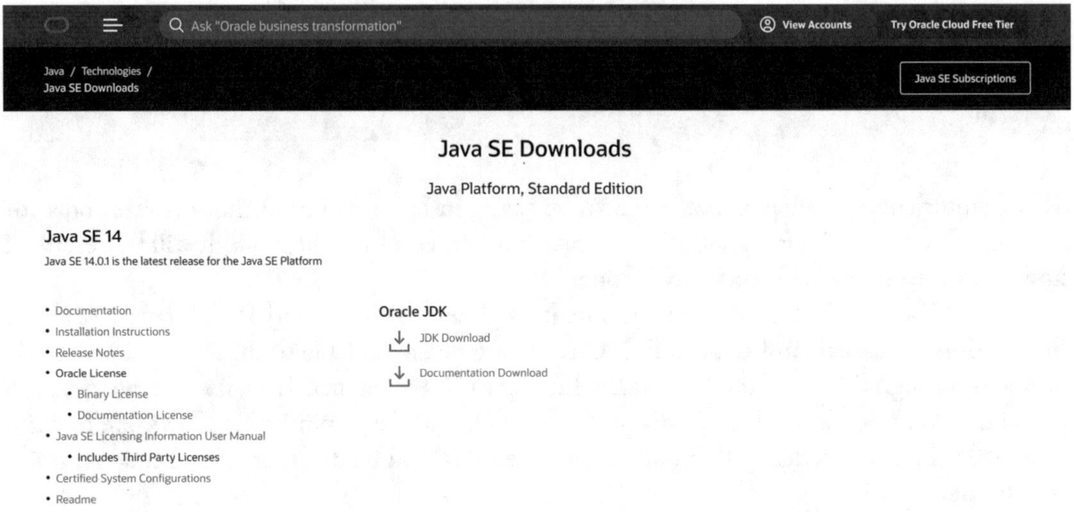

Figure 2.1: Java installation.

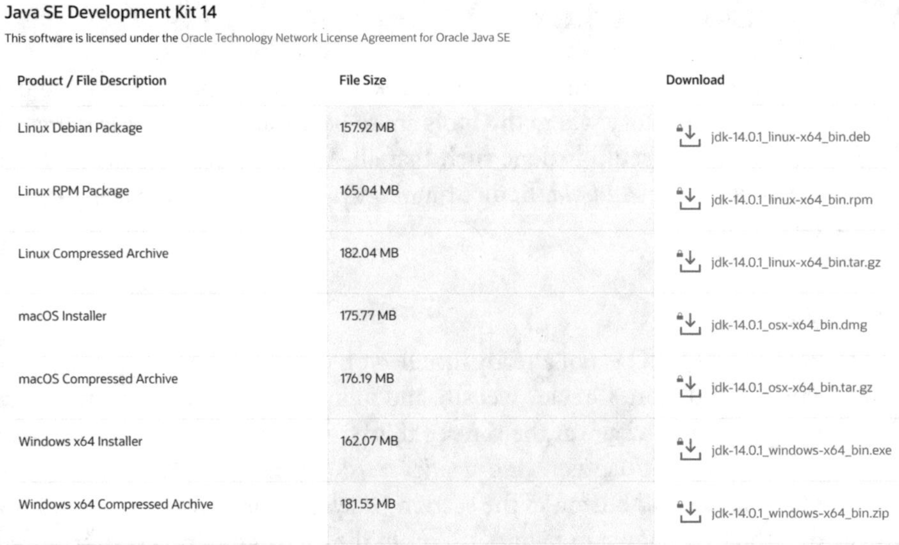

Java SE Development Kit 14

This software is licensed under the Oracle Technology Network License Agreement for Oracle Java SE

Product / File Description	File Size	Download
Linux Debian Package	157.92 MB	jdk-14.0.1_linux-x64_bin.deb
Linux RPM Package	165.04 MB	jdk-14.0.1_linux-x64_bin.rpm
Linux Compressed Archive	182.04 MB	jdk-14.0.1_linux-x64_bin.tar.gz
macOS Installer	175.77 MB	jdk-14.0.1_osx-x64_bin.dmg
macOS Compressed Archive	176.19 MB	jdk-14.0.1_osx-x64_bin.tar.gz
Windows x64 Installer	162.07 MB	jdk-14.0.1_windows-x64_bin.exe
Windows x64 Compressed Archive	181.53 MB	jdk-14.0.1_windows-x64_bin.zip

Figure 2.2: Java SE development kit 14.

android studio

Android Studio provides the fastest tools for building apps on every type of Android device.

DOWNLOAD ANDROID STUDIO

3.6.3 for Windows 64-bit (756 MB)

DOWNLOAD OPTIONS   RELEASE NOTES

Figure 2.3: SDK download path.

2.1.2   ANDROID STUDIO DEVELOPMENT ENVIRONMENT AND NATIVE DEVELOPMENT KIT

The most recent versions of Android Studio and the NDK at the time of this writing are used to run the lab experiments in the book. For the Windows, Mac, and Linux installation, the Android Studio is available as an executable installer which includes the development environment. For Android SDK tools, you need to download it separately. First, go the link https://developer.android.com/studio#downloads to see Figure 2.3. Then, click on the appropriate download option shown in this figure. A separate section of the same page gets opened. Find the section *Common Line Tools Only* and download the SDK for Windows platform. The NDK can be downloaded and installed from within Android Studio as described at http://developer.android.com/studio/index.html.

Download the Android Studio installation executable and run the Android Studio installer. For platform specific instructions, the installation instructions appear at https://developer.android.com/studio/install.html.

During the installation of Android Studio, there are two important settings that are critical to be performed correctly; see Figures 2.4 and 2.5. For the setting shown in Figure 2.4, make sure that all the components are selected for installation, and for the setting shown in Figure 2.5, make sure that Android Studio and Android SDK are installed in the directory *C:\Android*. To do so, manually create the directories by using the Browse option and create a *Studio* folder and an *sdk* folder. When the installer is finished, Android Studio can get started. Then, the Android Studio Setup Wizard is to be activated which is covered next. The Android NDK can be downloaded and installed from within Android Studio.

2.1.3   ANDROID STUDIO SETUP WIZARD

When Android Studio completes its installation, make sure the checkbox to run Android Studio is checked. The Android Studio Setup Wizard begins. Follow the steps noted as follows.

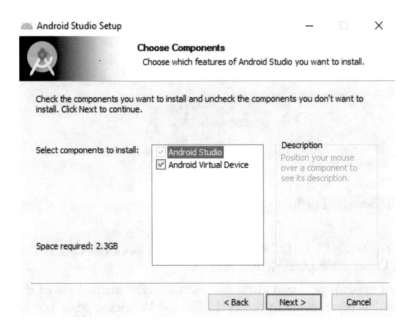

Figure 2.4: Android Studio setup.

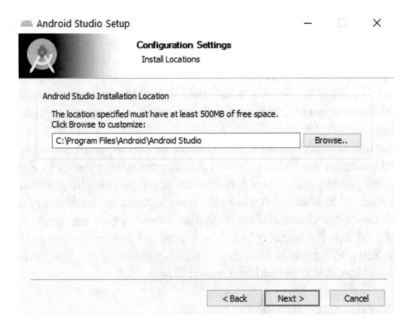

Figure 2.5: Configuration settings.

Figure 2.6: Custom installation.   Figure 2.7: SDK component setup.

- Hit next on the Welcome Screen.

- Select *Custom* to finish the configuration, as displayed in Figure 2.6.

- Select your preferred UI Theme *IntelliJ* or *Darcula*.

- Make sure all boxes are checked under Android SDK, Performance (HAXM), and Android Virtual Device (AVD).

  – Make sure to select the SDK Location as noted above, that is *C:\Android\sdk*.

- Use the recommended Emulator Settings.

- Review your settings and hit *Finish*.

When the above is done, the Android Studio home screen should appear, as shown in Figure 2.8.

Now run the *SDK Manager*, whose entry can be found by clicking on the *Configure* option. From this menu, additional system images for emulation and API packages for future Android versions can get added. Select any Android API level you may require listed on the SDK Platform tab; see Figure 2.9. On the SDK Tools tab, check the CMake, LLDB, NDK (Side by side) tools; see Figure 2.10. Click on *Apply* and follow the steps.

Allow the update process to complete.

2.1.4   ANDROID EMULATOR CONFIGURATION

The last item to take care of is configuring an AVD by clicking again on the Configure tab shown in Figure 2.8 and select the *AVD Manager* to open the AVD Manager shown in Figure 2.11. By default, Android Studio creates an x86 AVD. Since our development focus is ARM-based

Figure 2.8: Android Studio home screen.

Figure 2.9: SDK manager.

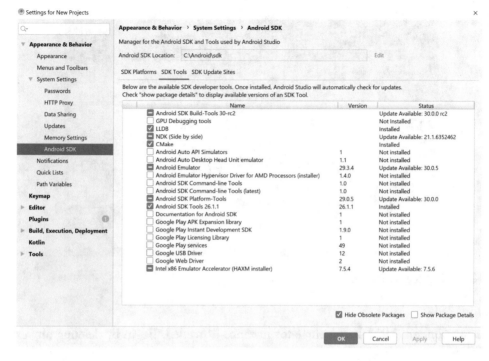

Figure 2.10: **SDK tools.**

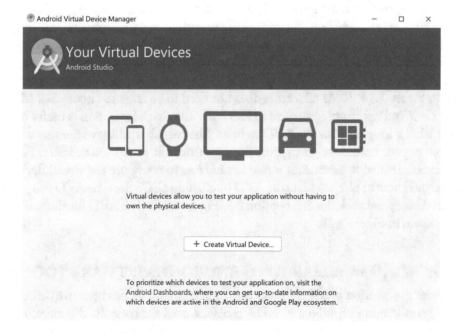

Figure 2.11: **Android virtual device.**

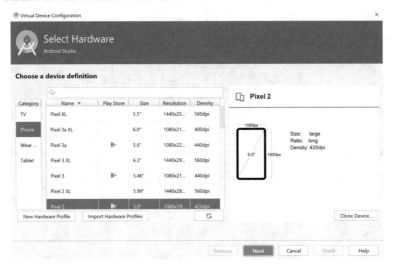

Figure 2.12: Device selection.

implementation, an ARM-based emulator instance is needed. Begin by deleting any existing AVD instances.

Click the button *Create Virtual Device...* to start configuring the AVD (see Figure 2.11). For the *Device* option shown in Figure 2.12, select a device with a good screen resolution for your computer—*Nexus S* is usually fine. For compatibility with newer smartphones, it is suggested to select the *Target* as the latest available version with the *CPU/ABI* as *ARM (armeabi-v7a)*. After selecting a proper device, you will be directed to *Verify Configuration* shown in Figure 2.13. Using the settings *Show Advanced Settings*, some changes can be made in the virtual device as illustrated in Figure 2.14. *RAM* allocation does not need to be large so choose 512 MB. Lastly, set the *SD Card* and *Internal Storage* size to 256 MB. The *Snapshot* option is useful to select as it normally takes a long time for the AVD to boot. The snapshot will save the memory state of the emulator to your hard drive so that starting the emulator occurs much faster. For the first time, it is recommended to select *Boot option* as *Cold boot* to configure the virtual device faster.

You should now be able to create the AVD by clicking *OK*. Select the AVD you just created in the list of devices and click the *Start* option (see Figure 2.15). The AVD for the created device appears as shown in Figure 2.16.

2.1.5   GETTING FAMILIAR WITH ANDROID SOFTWARE TOOLS

This lab covers the creation of a simple app on Android smartphones by constructing a "Hello World" program. Android Studio and NDK tools are used for code development, emulation, and code debugging. All the codes needed for this and other labs can be extracted from the book

Figure 2.13: Virtual device settings.

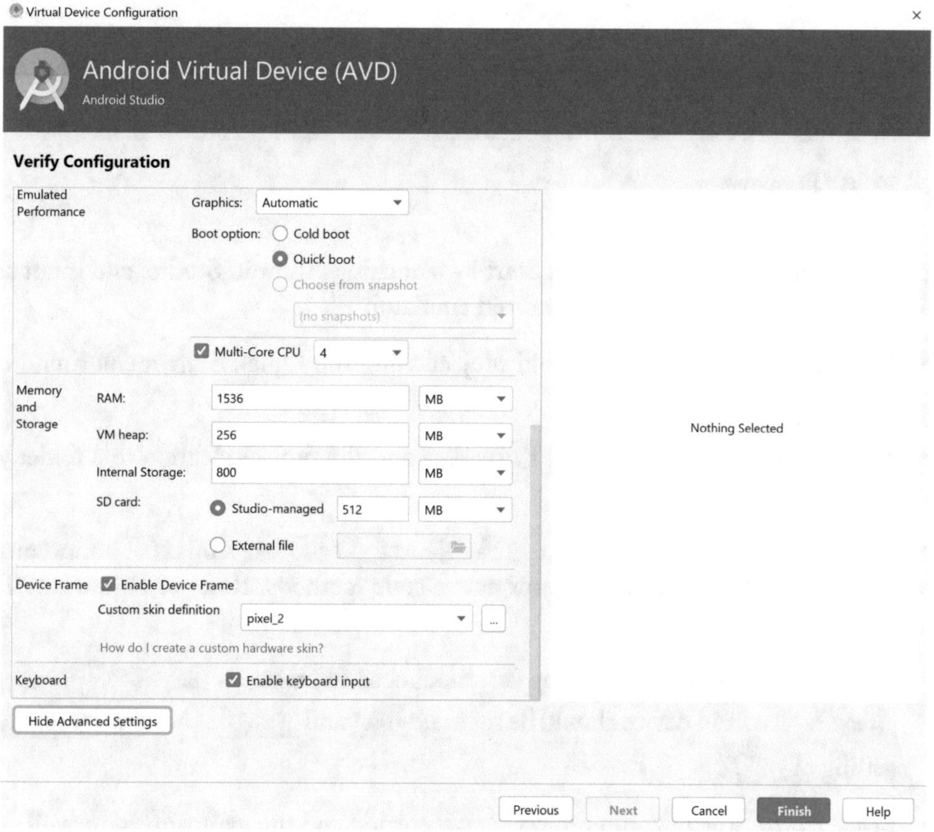

Figure 2.14: Advanced settings for virtual device configuration.

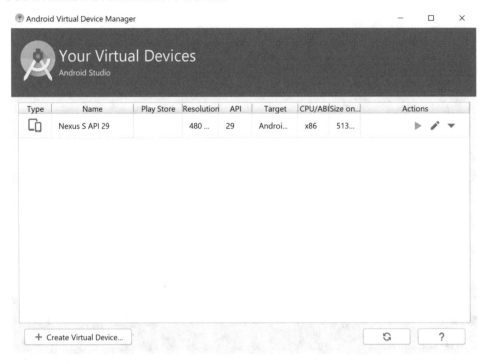

Figure 2.15: AVD manager.

codes package mentioned in the preface. Start by launching Android Studio, and if not already done, set up an AVD for use with the Android emulator.

- Begin by creating a new Android project using the Quick Start menu found on the Android Studio home screen.

- Set the *Application Name* to `HelloWorld` and the project location to a folder within the directory C:\Android.

- Change the *Package name* to `utdallas.edu.helloworld`. This is of importance later as it will affect the naming of your native code methods. Refer to Figure 2.17 for the previous three steps.

- Click *Next* and on the following screen, choose to create an `Empty Activity`. The `Target` Android device should be set to `Phone` and `Tablet` using a `Minimum SDK` setting of *API 15* .

- Select *Finish*. The new app project is now created and the main app editor will open to show the GUI layout of the app.

Figure 2.16: AVD appearance.

- Navigate to the *java* directory of the app in the *Project* window and open the *MainActivity.java* file under *utdallas.edu.helloworld*.

The entity that typically defines an Android app is called an *Activity*. Activities are generally used to define user interface elements. An Android app has activities containing various sections that users might interact with such as the main app window. Activities can also be used to construct and display other activities—such as if a settings window is needed. Whenever an Android app is opened, the onCreate function or method is called. This method can be regarded as the "main" (`C terminology`) of an activity. Other methods may also be called during various portions of the app lifecycle as detailed at the following website:

http://developer.android.com/training/basics/activity-lifecycle/starting.html

In the default code created by the SDK, `setContentView(R.layout.activity_main)` exhibits the GUI. The layout is described in the file *res/layout/activity_main.xml* in the *Package Explorer* window. Open this file to preview the user interface. Layouts can be modified using

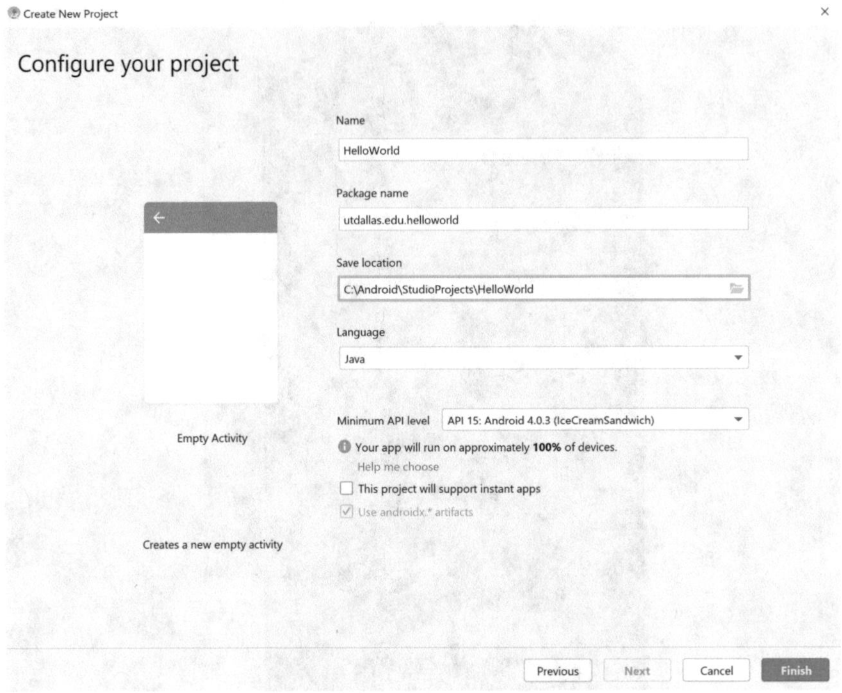

Figure 2.17: Initial screen of Android Studio.

the WYSIWYG editor which is built into Android Studio. For now the basic GUI suits our purposes with one minor modification noted below:

- Open the XML text of the layout (see Figure 2.18) by double clicking on the `Hello world!` text or by clicking on the *activity_main.xml* tab next to the *Graphical Layout* tab.

- Add the line `android:id ="@+id/Log"` within the `<TextView/>` section on a new line and save the changes. This gives a name to the TextView UI element.

TextView in the GUI acts similar to a console window. It displays text. Additional text can be appended by adding the `android:id` directive to the TextView code.

After setting up the emulator and the app GUI, let us cover interfacing with C codes. Note that it is not required to know the Java code syntax. The purpose is to show that the Java Native Interface (JNI) is a bridge between Java and C codes. Java is useful for handling Android APIs for sound and video i/o, whereas the processing codes are done in C. Note that to conduct the labs in this book, the programming is not done in C and C codes are generated by the MATLAB Coder discussed in the next chapter.

Figure 2.18: Hello World app.

A string returned from a C code is examined next. The procedure to integrate native code consists of creating a C code segment and performing alterations to the project. First, it is required to add support for the native C code to the project. The first step is to create a folder in which the C code will be stored. In the Project listing, navigate to *New → Folder → JNI* to create a folder in the listing called *jni*. Refer to Figures 2.19–2.22. Figure 2.19 shows how the Project listing view may be changed in order to show the *jni* folder in the main source listing.

Android Studio now needs to be configured to build a C code using the Gradle build utility. Begin by specifying the NDK location in the project *local.properties* file according to Figure 2.23. Assuming the directory *C:/Android* is used for setting up the development tools, the location specification would be as follows:

```
ndk.dir=C\:\\Android\\sdk\\ndk\\21.0.6113669
```

Next, the native library specification needs to get added to the *build.gradle* file within the project listing under the "app" folder. This specification declares the type of external build and its configuration file which defines the name of the external library that Java will load. This is done by adding the following code within the *android* section:

```
externalNativeBuild  {
    cmake {
        path "CMakeLists.txt"
    }
}
```

Figure 2.19: **JNI folder.**

Figure 2.20: **New Android activity.**

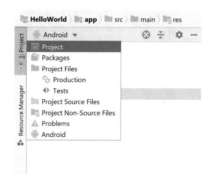

Figure 2.21: **Project listing.**

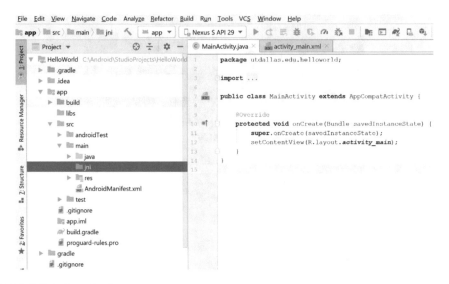

Figure 2.22: **MainActivity.java.**

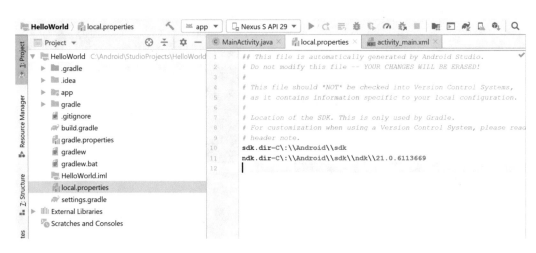

Figure 2.23: local.properties.

The correct placement of the code is highlighted in Figure 2.24. Another part that needs to get added is a `CMakeLists.txt` file in the *project* section and app folder as shown in Figure 2.25 and by having the following code inside the .txt file:

cmake_minimum_required (**VERSION 3.4.1**)
 # Creates and names a library, sets it as either STATIC
 # or SHARED, and provides the relative paths to its source code.
 # You can define multiple libraries, and CMake builds them for you.
 # Gradle automatically packages shared libraries with your APK.
add_library (*# Sets the name of the library.*
            HelloWorld
            # Sets the library as a shared library.
            SHARED
            # Provides a relative path to your source file(s).
            src/main/jni/HelloWorld.c
)

The C code considered here consists of a simple method to return a string when it is called from the `onCreate` method. First, the code that defines the native method needs to be included. Create a new *HelloWorld.c* file. Add the following code and save the changes:

```
#import <jni.h>
jstring Java_utdallas_edu_helloworld_MainActivity_getString
( JNIEnv* env, jobject thiz )  {
      return (*env)->NewStringUTF(env, "Hello UTD!");
}
```

This code defines a method that returns a Java string object according to the JNI specifications with the text `Hello UTD!` The naming for this method is dependent on what is called *fully qualified name* of the native method which is defined in *MainActivity*. There are alternate methods of defining native methods that will be discussed in later labs.

Next, the native method needs to be declared within *MainActivity.java* (see Figure 2.26) according to the naming used in the C code. To do so, add this declaration below the `onCreate` method already defined.

```
public native String getString();
```

Then, add the following code within `public class` to load the native library:

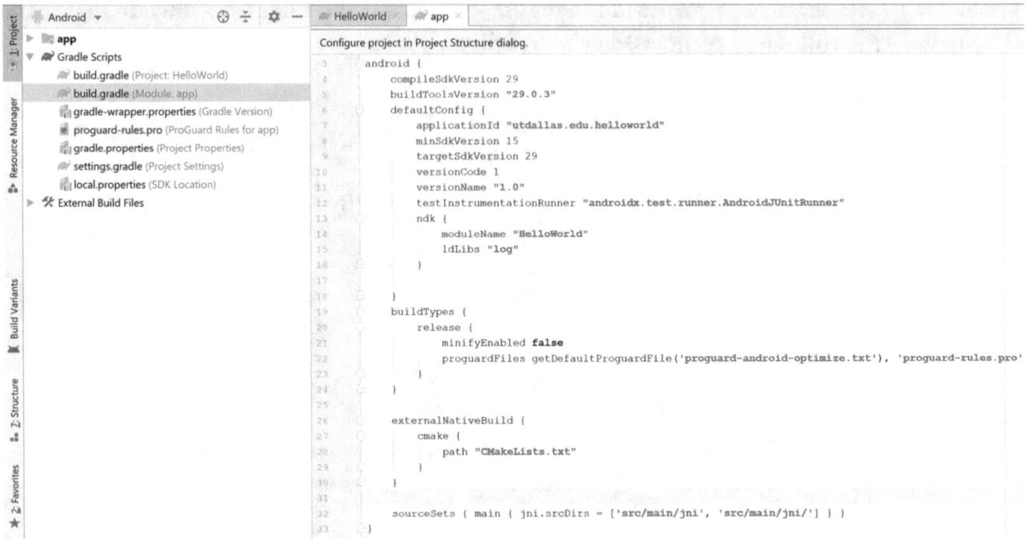

Figure 2.24: **Code placement.**

Figure 2.25: **CMakeLists placement.**

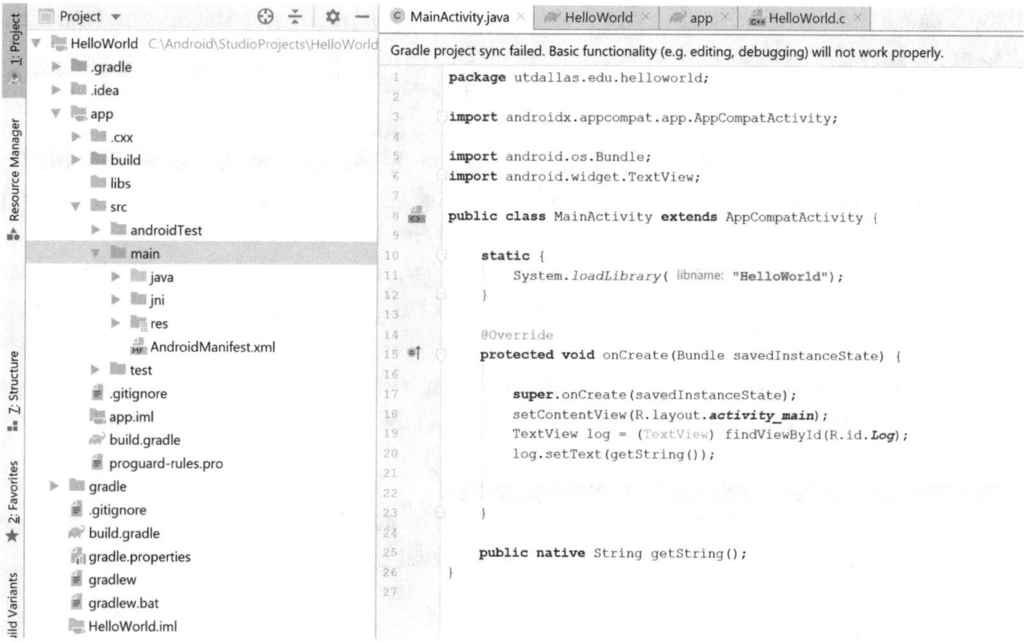

Figure 2.26: TextView.

```
static {
    System.loadLibrary("HelloWorld");
}
```

To use the TextView GUI object, it needs to be imported by adding the following declaration to the top of the *MainActivity.java* file:

```
import android.widget.TextView;
```

The TextView defined in the GUI layout needs to be hooked to the `onCreate` method by adding the following lines to the end of the `onCreate` method code section (after `setContentView` but inside the bracket):

```
TextView log = (TextView)findViewById(R.id.Log);
log.setText( getString() );
```

This will cause the text displayed in the TextView to be changed by the second line which calls the C `getString` method.

Save the changes and select the *Make Project* option (located under the Build category on the main toolbar). Android Studio would display the build progress and notify if any errors occur. Next, run the app on the Android emulator using the *Run app* option located in the Run menu of the toolbar. If an emulator is already running, an option will be given to deploy the app to the selected device (see Figure 2.27). Android Studio should launch the emulator and the screen (see Figure 2.28) would display `Hello UTD!` . To confirm that the display is being changed, comment out the line `log.setText()` and run the app again. This time the screen would display `Hello World!` .

Note that the LogCat is equivalent to the main system log or display of the execution information. Here, the code from the previous project is modified to enable the log output capability as follows.

- Add the logging library to the *build.gradle* file by adding the line `ldLibs "log "` (see Figure 2.24) to the *ndk* section which was added previously.

- Add the Android logging import to the top of the *HelloWorld.c* source file by adding the line `#include <android/log.h>` .

- Add the following code to output the test message before the `return` statement:

```
int classNum = 9001;
int secNum = 1;
__android_log_print(ANDROID_LOG_ERROR, "HelloWorld",
"DSP %d.%03d", classNum, secNum);
```

The android_log_print() method (two underscores at the beginning) is similar to the `printf function in C` . The first two parameters are the log level and the message tag. In the above example, the string has a specified integer for the class number inserted, followed by a specified integer for the section number. The same number formatting that is possible when using the `printf` function may also be used here. For instance, the section number can be formatted to three characters width with leading zeros. Variables are last and are inserted with the formatting specified in the message string in the order they are listed.

Save the changes made to the *HelloWorld.c* source file and run the app again. This time, Android Studio should automatically open the Android DDMS window and show the LogCat screen. The message `DSP 9001.001` would appear in the listing if the previous procedures were performed properly (see Figure 2.29).

Figure 2.27: Choose device.

Figure 2.28: App screen.

Figure 2.29: LogCat.

2.2   iOS TOOLS INSTALLATION STEPS

Similar to the previous section, this section covers the installation of the necessary software packages for iOS app development. This time a "Hello World!" app in the iOS development environment is constructed.

C code segments are made available to the iOS Objective-C environment through the normal header files that are used in C. Objective-C allows C codes to run without the need for any external wrapper. For accessing inputs and outputs or sensor signals on iPhones, the publicly available iOS APIs in Objective-C are used.

The development environment consists of the XcodeIDE. This development environment allows writing C codes, compiling, and debugging on an iOS device simulator or on an actual iOS device. The Xcode IDE includes a built-in debugger that can be used to debug C codes line-by-line and also to observe values stored for different variables. Xcode is available as a free download on Mac machines through the Apple App Store.

To develop iOS apps, the following items are needed:

- an Apple Mac computer,

- enrollment in an Apple-approved developer program, and

- an iOS device.

Note that in the absence of an actual iOS device, the iOS Simulator can still be used. Different iOS configurations can be selected from the scheme selector, which is located at the top left of the Xcode window.

2.2.1   iPHONE APP DEVELOPMENT WITH XCODE

1. Launch the Xcode IDE. You should be prompted with a splash screen as shown in Figure 2.30. Select *Create a new Xcode project*. In case this screen does not get displayed, you can use *File -> New -> Project*.

2. Select *iOS -> Application -> Single View Application* (see Figure 2.31). After clicking *Next*, the configuration of the project appears as shown in Figure 2.32.

3. Enter *Product Name* as HelloWorld.

4. Enter *Organization Name*.

5. Enter an *Organization Identifier*.

6. Set *Language* to Objetive-C and *Devices* to iPhone.

7. Leave deselected Use Core Data, Include Unit Tests, and Include UI Tests.

8. Click *Next*. On the next page, remember to deselect *Create Git Repository on*.

9. Select the destination to store your project and select *Create*.

After clicking *Create*, the settings screen of the project gets shown. Here the features of the app can be altered. The devices supported by your project can be changed and also any additional frameworks or libraries to be utilized by your project can be added.

If getting a warning display "No signing identity found," this means you need to have your Apple Developer Account accepted for iOS app development. Also, your device must be certified for app development.

2.2.2   SETTING-UP APP ENVIRONMENT

The left column in the Xcode window is called the Navigator. Here one can select or organize different files and environment for a project.

- In the Navigator Pane, the Main.Storyboard entry is seen. This is used to design the layout of your app. Different UI elements in multiple views provided by the IDE can be used to design the interface of an app. However, this is done programmatically for this example.

- AppDelegate.h and AppDelegate.m are Objective-C files that can be used to handle events such as:

  - app termination,
  - app entering background or foreground, and

Welcome to Xcode

Version 9.2 (9C40b)

Get started with a playground
Explore new ideas quickly and easily.

Create a new Xcode project
Create an app for iPhone, iPad, Mac, Apple Watch or Apple TV.

Clone an existing project
Start working on something from an SCM repository.

Figure 2.30: Xcode welcome screen.

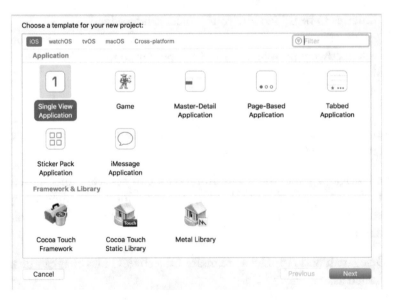

Figure 2.31: Xcode new project selection.

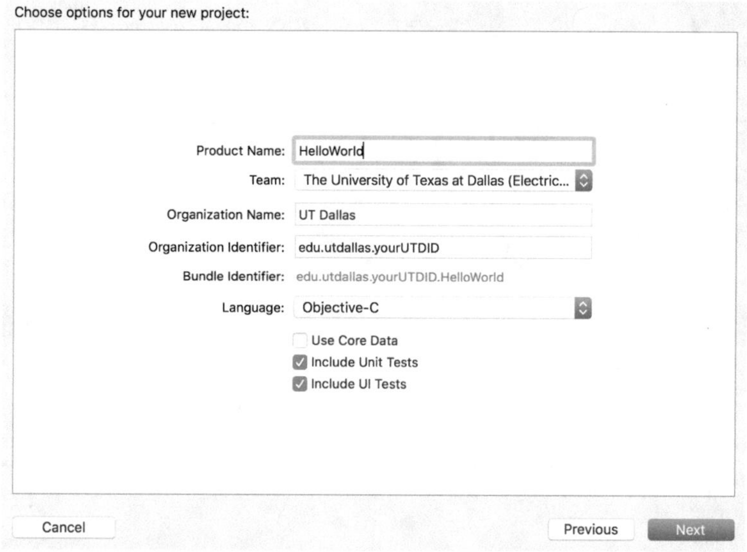

Choose options for your new project:

Product Name:	HelloWorld
Team:	The University of Texas at Dallas (Electric...
Organization Name:	UT Dallas
Organization Identifier:	edu.utdallas.yourUTDID
Bundle Identifier:	edu.utdallas.yourUTDID.HelloWorld
Language:	Objective-C

☐ Use Core Data
☑ Include Unit Tests
☑ Include UI Tests

Cancel   Previous   Next

Figure 2.32: Xcode project options.

– app loading.

These files are not accessed here.

• The files ViewController.m and ViewController.h are used to define methods and properties specific to a particular view in the storyboard.

2.2.3 CREATING LAYOUT

In the file ViewController.m, the method called `viewDidLoad` is used to perform processes after the view has loaded successfully. This method is used here to initialize the UI elements.

In the interface section of `ViewController`, add the following two properties: a label and a button:

```
@interface ViewController()
@property UILabel *label;
@property UIButton *button;
@end
```

Initialize the label and button. Then, assign them to the view. This can be done by adding this code in the method `viewDidLoad`:

```
_label = [[UILabel alloc] initWithFrame:CGRectMake( 10,15,300, 30)];
_label.text = @"Hello World!";
[self.view addSubview:_label];
_button = [UIButton buttonWithType:UIButtonTypeRoundedRect];
_button.frame = CGRectMake(10, 50, 300, 30);
[_button setTitle:@"Click" forState:UIControlStateNormal];
[self.view addSubView:_button];
[_button addTarget:self action:@selector(buttonPress:)
                      forControlEvents:UIControlEventTouchUpInside];
```

An action is attached to the button `buttonPress` . This action will call the method which gets executed when the button is pressed. As the control event is UIControlEventTouchUpInside, the method gets executed when the user releases the button after pressing. This method is declared as follows:

```
(IBAction)buttonPress:(id)sender {
}
```

As of now, it is a blank method. A property is assigned to it after specifying the C code that is to be used to perform a signal processing function.

The app can be run by pressing the *Play* button on the top left of the Xcode window. An actual target is not needed here and one my select the simulator; see Figure 2.33. For example, iPhone 6 can be selected as the simulator.

When the app is run, the label "Hello World!" can be seen and the button gets created. On clicking the button, nothing happens. This is because the method to handle the button press is empty.

It is not required to know the Objective-C syntax. Note that one can call a C function in Objective-C just by including a header. Objective-C is useful for handling iOS APIs for sound and video i/o, whereas signal processing codes are done here in C.

2.2.4   IMPLEMENTING C CODES

In this section, a C code is linked to ViewController using a header file.

- Right click on the HelloWorld folder in your project navigator in the left column and select *New File*.

- Select iOS -> Source -> C File.

- Write the filename as Algorithm and select *Also create a header file*.

Figure 2.33: Select iPhone Simulator.

- After clicking *Next*, select the destination to store the files. Preferably store the files in the folder of your project.

- In the project navigator, you can view the two new added files. Select Algorithm.c.

- In Algorithm.c, enter the following C code:

```
const char *HelloWorld()  {
      printf("Method called\n");
      return "Hello UTD!";
}
```

- The function `HellowWorld()` prints a string and returns a char pointer upon execution. Let us call this function on the button press action in the view controller and alter the label.

- To allow this function to be called in Objectve-C, the function in the header file needs to be declared. For this purpose, in Algorithm.h, add the following line before `#endif` :

```
const char *HelloWorld();
```

- Now a C function is created, which is called and executed via Objective-C, just by including the header file.

2.2.5   EXECUTING C CODES VIA OBJECTIVE-C

Now that a C code is written, it needs to be linked to the Objective-C app in order to be executed.

- In the ViewController.m, just below `#import ''ViewController.h''`, add `#import ''Algorithm.h''` .

- In the `buttonPress` method, include the following line:

```
_label.text = [NSString stringWithUTF8String:HelloWorld()];
```

This code line alters the text of the label in the program.

- Run the program using the simulator.

  1. The text of the label changes.
  2. In the Xcode window, "Method Called" gets printed in the Debug Console at the bottom.

This shows that printing can be done from the C function to the debug console in Xcode. This feature is used for debugging purposes.

2.2.6   iOS APP DEBUGGING

After getting some familiarity with the Xcode IDE by creating and modifying an iOS app project and running the app on an iPhone simulator, the following steps indicate how to debug C codes via the built-in Xcode debugger.

To obtain familiarity with the Xcode debugging tool, perform the following.

- Open the HelloWorld project.

- Select Product -> Build.

- In the project navigator, select the Algorithm.c file.

```
 9  #include "Algorithm.h"
10
11  const char *HelloWorld() {
12      printf("Method Called\n");
13      return "Hello UTD!";
14  }
15
```

Figure 2.34: A debug point added on line 12.

After the project is successfully built, so-called debug points can be placed inside the C code. Debug points can be placed by clicking on the column next to the line to be debugged or by pressing CMD + \ . A blue arrow appears, see Figure 2.34, that points toward the line to be debugged.

The Xcode debugger allows one to:

- pause the execution at a particular line of code,

- know the value of the variable at that particular instant of execution, and

- navigate from the function call to function execution as it gets executed.

CHAPTER 3

From MATLAB Coder to Smartphone

This chapter presents the steps one needs to take in order to run signals and systems algorithms written in MATLAB on the ARM processor of smartphones. The steps needed are best conveyed by going through an example. This example involves a simple signal generation algorithm. The same steps can be used to run other MATLAB codes that have already been written and are publicly available. An alternative approach based on the use of Simulink Coder was reported in [1].

3.1   MATLAB FUNCTION DESIGN

This section provides the steps needed in order to run a simple signal generation algorithm written as a MATLAB script on smartphones. Before going through the example, the difference between a MATLAB function and a MATLAB script is first stated and it is shown how to convert a MATLAB script to a MATLAB function.

In Chapter 1, it was shown that a MATLAB script reads commands and executes them as they are typed sequentially. In general, a program can be a script that executes a series of MATLAB commands, or can be a function that accepts inputs and generates outputs. Both scripts and functions contain a series of MATLAB commands which are written as text and saved in .m format, however MATLAB functions provide more flexibility.

Let us start with a MATLAB script. Create a script, *New → Script*, from the HOME panel. Write the following code to evaluate the expression $a^3 + \sqrt{bd} - 4c$:

```
>> a=2;
>> b=3;
>> c=4;
>> d=5;
>> a^3+sqrt(b*d)-4*c
```

Save the script by specifying a desired name, for example *MyEquation*. Run the code. For the values indicated in the Command Window, this response appears:

```
ans =
-4.1270
```

To accommodate for different *a*, *b*, *c*, and *d* values, instead of manually changing them in the script and seeing the result, the script can be written as a function. To create a function, open a new script and state a list of the outputs, the function name, and a list of the inputs as follows:

```
function  [output1, output2,...] =
          name of the function(input1, input2,...)
```

Then, enter it as a MATLAB command after its definition. The input parameters of the function are the parameters which control the outputs. After entering the script, save it. You will see that MATLAB recognizes the name as you have defined it for the function (MATLAB function filenames need to have the .m extension). Upon acceptance, save it in .m format; see Figure 3.1.

After saving this MATLAB function, to see the result for different *a*, *b*, *c*, and *d* values, you now only need to write the function with the desired input values in the Command Window, that is:

```
Result = MyEquation(2,3,0,6)
```

The created MATLAB function can be used similar to other MATLAB functions in any program. In practice, when an algorithm involves different parts, it is a good programming practice to have different parts of the algorithm in separate MATLAB functions instead of having all the parts written in one script.

3.2   GENERATING SIGNALS VIA MATLAB ON SMARTPHONES

This section covers the steps for generating a pulse and an exponential signal, see Figure 3.2, on a smartphone. The first step involves opening the MATLAB environment and creating a new function file for the algorithm.

Let us name the function Lab3_1. Enter the following code. In the MATLAB code, need to keep the order of the input parameters the same as originally defined.

Note that throughout the book, the notations dt, delta, and Δ are used interchangeably to denote the time interval between samples.

Figure 3.1: MATLAB function.

```
function [x1,x2]=Lab3_1(a,b,Delta)
t=0:Delta:8-Delta;      % Time at which samples are generated
x1=a*[ones(1,4/Delta) zeros(1,4/Delta)];      %Pulse function
x2=exp(-b*t);           %exponential function
```

This function has three inputs, `a`, `b`, and `Delta`. The parameter `a` controls the amplitude of the pulse signal and the parameter `b` controls the decreasing rate of the exponential function. The parameter `Delta` controls the time resolution or spacing between time samples. `x1` and `x2` denote the pulse and the exponential signals, respectively.

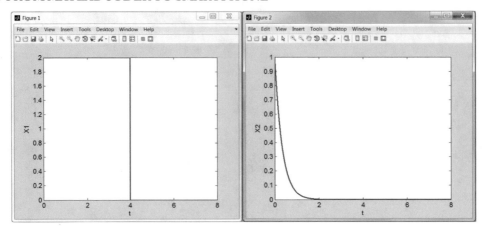

Figure 3.2: Signal generation.

3.2.1   TEST BENCH

A useful next step is functional verification. This step helps verifying the response on a smart-phone target. The following script shows how such verification is achieved for the above algorithm:

```
clear;
clc;
Delta=0.001;  %Time interval between samples
a=2;          %amplitude of the pulse signal
b=3;          %Decay rate of the exponential function
              %Calling the function which performs operations
[x1,x2]=Lab3_1(a,b,Delta);
```

This script generates the two signals corresponding to a= 2 and b = 3 for Delta = 0.001 .

3.2.2   C CODE GENERATION

After defining the MATLAB function and the test bench script code, an equivalent C code needs to be generated via the MATLAB Coder. The MATLAB Coder (or simply Coder) can be found in the toolbar under APPS. The following screenshot shows the process of generating an equivalent C source code from a MATLAB code. Before performing this step, it is important to make sure that the directory of the MATLAB files appears as the MATLAB directory. Run the test bench code and then take the steps stated in Figure 3.3.

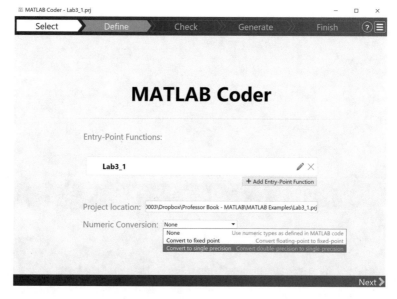

Figure 3.3: MATLAB Coder function selection.

Figure 3.3 shows the initial screen displayed by the Coder. Start by selecting the function to be converted and change the *Numeric Conversion* option to *single precision floating-point arithmetic*. Note that the *single precision floating-point arithmetic* is available in MATLAB 2019b and later versions and care must be taken in using the right version of MATLAB.

After the function is selected, the input types need to be specified (see Figure 3.4). This can be done either manually or by using the above test bench script to automatically determine the data types. Select the test bench script and choose the *Autodefine Input Types* option to complete this step.

After the input types are set, the Coder then checks to ensure that it is able to generate a C code from the provided MATLAB script. Figure 3.5 shows the outcome with no detected errors. Although many of the built-in MATLAB functions work with the MATLAB Coder, not all the functions are supported. Unsupported functions need to be written from scratch by the programmer.

Once the MATLAB script is checked and passed, a corresponding C source code can be generated by pressing the *Generate* button (see Figure 3.6). Although various configuration settings are available, the default settings are adequate for our purposes. After this step, a folder named *codegen* is created in the directory of the MATLAB files.

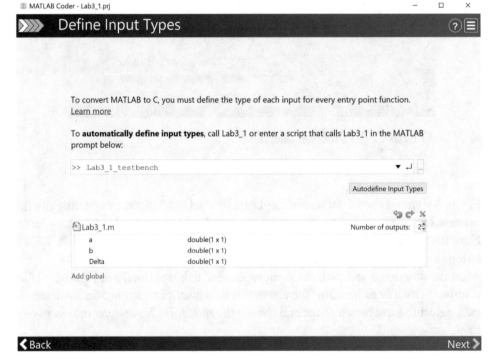

Figure 3.4: **Input types.**

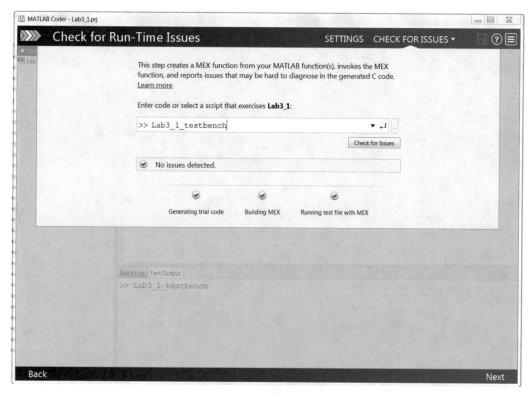

Figure 3.5: Function error check.

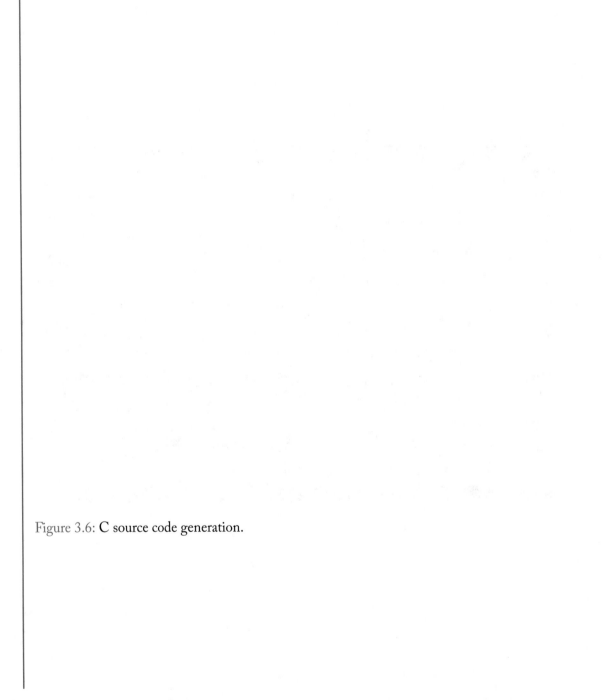

Figure 3.6: C source code generation.

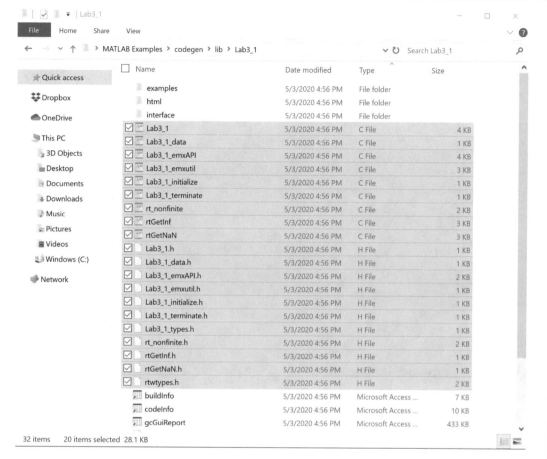

Figure 3.7: **Copying files.**

3.2.3   SOURCE CODE INTEGRATION

The final step to implement an algorithm written in MATLAB on smartphones is to deploy the generated C code on a suitable target device. For this step, an application shell is provided here in which the generated C source code needs to be placed. This shell operates in the same manner as the test bench script stated earlier, that is the two signals (pulse and exponential) are generated. Figures 3.7 and 3.8 show how the generated C code is placed or integrated into the Android shell that is provided. The following steps need to be taken.

1. Navigate to folder named Lab3_1 (MATLAB function name) in the *codegen* folder; *codegen/lib/Lab3_1*. Copy all the files with .h and .c extensions; see Figure 3.7.

2. Place the copied files inside the *jni* folder of the shell provided. The *jni* folder appears at *app/src/main/jni*; see Figure 3.8.

Figure 3.8: C source code integration.

3. If using MATLAB R2016b or later, it is required to add a specific header file into the *jni* folder separately. This header file (`tmwtypes.h`) can be found in the MATLAB root with the following path:

```
MATLAB root\R2019b\extern\include
```

4. Before building the project, first press Clean Project, as shown in Figure 3.9.

The MATLAB Coder produces C codes with the required include statements and function calls for using them. One minor modification of a generated C code that may be required is to ensure having the correct input and output variable data types. In case of array inputs, the generated code is specified using static array sizes and thus needs to be modified to access array pointers.

Finally, it is worth stating that the most important consideration when transiting a MATLAB function to smartphones is awareness of input and output data types. A persistent variable storage needs to be established by declaring a persistent variable and performing a one-time initialization. After this declaration and initialization, any data may be retained between calls to

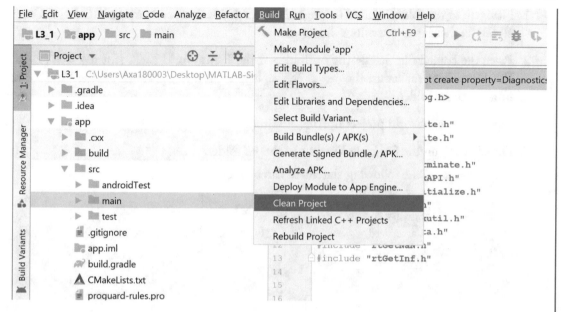

Figure 3.9: Clean Project.

the function. With the approach presented in this chapter, practically any signals and systems algorithm written in MATLAB can be made to run on smartphones.

3.3   RUNNING MATLAB CODER-GENERATED C CODES ON SMARTPHONES

3.3.1   RUNNING ON ANDROID SMARTPHONES

This section covers the steps for integrating a C code generated by the MATLAB Coder into an Android shell program for running it on an Android smartphone. The steps are as follows.

1. Creating a shell.

2. Verification of the MATLAB function to be run.

3. Using the MATLAB Coder to generate the corresponding C code to be placed in the shell.

4. Modifications of the shell to integrate the C code into it.

The first three steps were covered in the previous sections. In this section, the fourth step is covered. Let us consider the basic HelloWorld shell that was mentioned and the corresponding C code generated by the MATLAB Coder. The following modifications of the shell are needed for running the code on an Android smartphone. The HelloWorld shell for Lab3_1 is used here.

1. Navigate to the folder created by the MATLAB Coder named *codegen\lib\Lab3_1*. Copy all the files with the extensions .h and .c.

2. Place the copied files inside the *jni* folder of the HelloWorld shell. If using MATLAB R2016b or later, make sure the `tmwtypes.h` header file is also placed from the MATLAB root in the *jni* folder.

3. Open the HelloWorld project in Android Studio.

4. Double click on the file HelloWorld.c in the *jni* folder.

   Inside the code, the following statement can be seen:

```
#import <jni.h>
jstring Java_com_dsp_helloworld_MainActivity_getString
( JNIEnv* env, jobject thiz ) {
        return (*env)->NewStringUTF(env, "Hello UTD!");
}
```

5. Go to the folder *codegen\lib\Lab3_1\examples* and open the file main.c using Notepad.

6. Copy the part on the top specified by Include Files as shown below, except main.h, into the HelloWorld.c file.

7. Next, modify the statement line from:

```
jstring Java_com_dsp_helloworld_MainActivity_getString
( JNIEnv* env, jobject thiz )
```

   to

```
jdoubleArray Java_com_dsp_helloworld_MainActivity_getStringx1 (
        JNIEnv* env, jobject thiz, jdouble a, jdouble b,
        jdouble delta )
```

   Note `jdoubleArray` is used instead of `jstring` because the intention in the HelloWorld project was to return the string "Hello UTD" but here the intention is to return an array of double type numbers. Add the input parameters and their data types of the MATLAB function here. In Lab3_1, these parameters are *a*, *b*, and *delta*, that is:

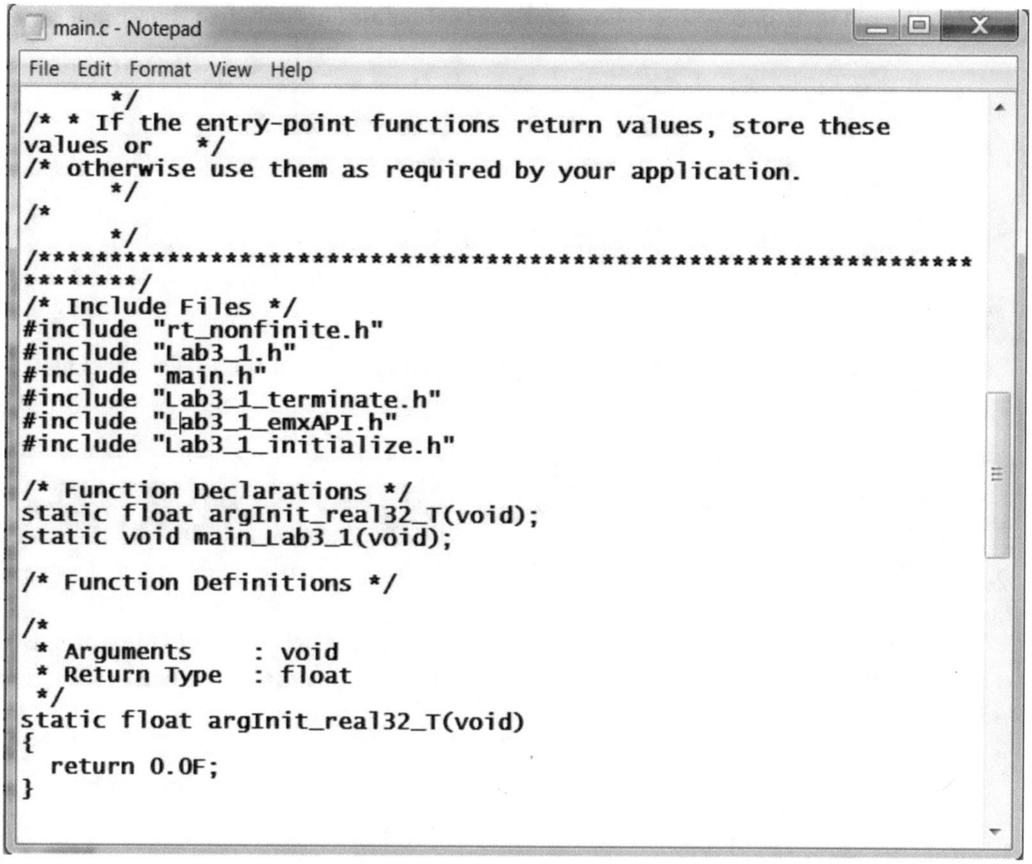

```
main.c - Notepad

File  Edit  Format  View  Help
     */
/* * If the entry-point functions return values, store these
values or     */
/* otherwise use them as required by your application.
         */
/*
        */
/****************************************************************
********/
/* Include Files */
#include "rt_nonfinite.h"
#include "Lab3_1.h"
#include "main.h"
#include "Lab3_1_terminate.h"
#include "Lab3_1_emxAPI.h"
#include "Lab3_1_initialize.h"

/* Function Declarations */
static float argInit_real32_T(void);
static void main_Lab3_1(void);

/* Function Definitions */

/*
 * Arguments      : void
 * Return Type   : float
 */
static float argInit_real32_T(void)
{
  return 0.0F;
}
```

Figure 3.10: main.c file located in the folder codegen\.lib\Lab3_1\examples.

```
jdoubleArray Java_com_dsp_helloworld_MainActivity_getString (
        JNIEnv* env, jobject thiz, jdouble a, jdouble b,
        jdouble delta )
```

8. Then, inside the function created, place the following statements:

```
int length=round(8/delta);
jdoubleArray jArray = (*env)->NewDoubleArray(env, length);
```

These statements define the output variable and allocate memory to it.

From the file main.c, copy the following lines that define and initialize the variables:

```
    /* Initialize function 'Lab3_1' input arguments. */
emxArray_real32_T *x1;
emxArray_real32_T *x2;
emxInitArray_real32_T(&x1, 2);
emxInitArray_real32_T(&x2, 2);
```

Now copy the following line from the file main.c and place it in the HelloWorld file:

```
/* Call the entry-point 'Lab3_1'. */
Lab3_1(argInit_real32_T(), argInit_real32_T(), argInit_real32_T(),
x1, x2);
```

This function calls the function Lab3_1, where two output arrays x1, x2 are returned. To assign x1 data values to `Array` , a temporary array called `yac_temp` is used within a for loop as follows:

```
int i;
double yac_temp[length];
for( i = 0; i < length; i++) {
    yac_temp[i] = x1->data[i];
    // __android_log_print(ANDROID_LOG_ERROR,
    //           "HelloWorld Final 1", "%f\n", y_temp[i]);
}
```

Next, free the allocated memory by copying the following lines from the file main.c:

```
emxDestroyArray_real32_T(x2);
emxDestroyArray_real32_T(x1);
```

Assign `yac_temp` to `Array` and return it:

```
(*env)->SetDoubleArrayRegion(env,jArray,0,length,yac_temp);
return jArray;
}
```

These steps need to be repeated for all the defined outputs in the function Lab3_1; see the shell provided for Lab3_1.

9. In the Project Navigator, open the file CMakeLists.txt. Under the comment "#Provides a relative path to your source files(s)", enter the filenames of all the .c and .h files from MATLAB. Make sure to include the `tmwtypes.h` header into the CMakeLists.txt as well. This will let Android Studio and CMAKE know of these files.

10. Finally, if desired, design your own Graphical User Interface (GUI). This link provides guidelines as how to design GUIs: http://androidplot.com/docs/quickstart/. An example GUI designed for this application is shown in Figure 3.11.

3.3.2   RUNNING ON iOS SMARTPHONES

This section covers the steps for integrating a C code generated by the MATLAB Coder into an Xcode shell program for running it on iPhone smartphones. These steps are listed below.

1. Creating a shell.

2. Verification of the MATLAB function to be run.

3. Using the MATLAB Coder to generate the corresponding C code to be placed in the shell.

4. Modifications of the shell to integrate the C code into it.

The first two steps were covered previously in the Android section and are the same for iOS. In this section, the third and fourth steps are covered. Let us consider the lab L3_1 shell and its corresponding C code generated by the MATLAB Coder. The following modifications of the shell are needed for running the code on an iOS smartphone.

Here it is worth mentioning that the group structure in Xcode is not the same as the folder structure in Finder on Mac. It can be made to be the same, but making a group does not make a relative folder on the file system.

1. Make the directory codegen/lib/L3_1 as a subfolder in the Xcode project L3_1.

2. Navigate to the folder created by the MATLAB Coder named *codegen\lib\L3_1*. Copy all the files with the extensions .h and .c to the Xcode project. Note that if using MATLAB

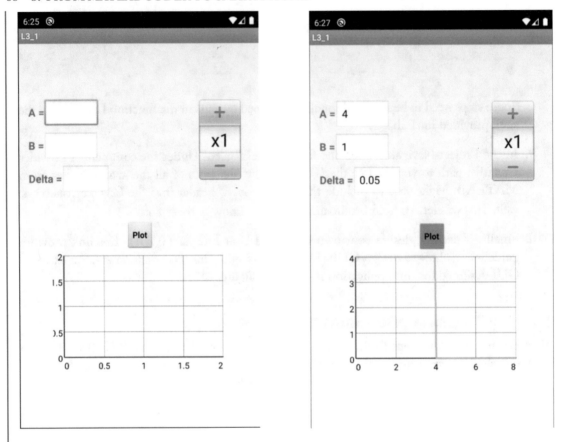

Figure 3.11: Initial screen of Lab3_1 app and plot of x1.

R2016b or later, it is required to copy a specific header file separately. This header file (`tmwtypes.h`) can be found in the MATLAB root with the following path:

```
MATLAB root\R2019b\extern\include
```

3. In the Xcode Project Navigator, click on *File -> Add Files to "L3_1"*.

4. Click File -> New -> Group and type in "Native Code".

5. Select all of the .c and .h files and drag them into the "*Native Code*" group.

The MATLAB C codes will be built and get ready for use. The code to interact with buttons and graphs was covered earlier. Now, the generated C code needs to get placed or integrated into the iOS shell.

Click on "ViewController.m" to edit the file. At the top of the file above the interface definition, import the following necessary header files:

```
#import "ViewController.h"
#import "Lab3_1.h"
#import "Lab3_1_emxAPI.h"
```

Underneath the line, in the buttonPress method,

```
int length = round( 8.0 / delta);
```

add the following lines:

```
emxArray_real32_T *x1 = emxCreate_real32_T(1,length);
emxArray_real32_T *x2 = emxCreate_real32_T(1,length);
emxArray_real32_T *data = x1;
Lab_3_1(a, b, delta, x1, x2);

switch( _selection )
{
    case 0:
        data = x1;
        break;
    case 1:
        data = x2;
        break;
}

[_graphView updateBuffer:data->data withBufferSize:length];

emxDestroyArray_real32_T(x1);
emxDestroyArray_real32_T(x2);
```

The first two lines allocate memory for output data x1 and x2. The variable "data" is used as a placeholder to refer to the data to be plotted. The switch statement picks the section of the

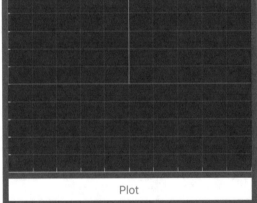

Figure 3.12: Plot using EZPlot for L3_1 on iOS platform.

code that will assign the data to be plotted. The next line updates the graph object, providing the location of the data and its length. Last, the memory for the two outputs are deallocated.

The iPhone apps use Cocoa Pods which are essentially a bundle or a library of codes. The Pods used here in the Xcode shell projects are CorePlot and EZAudio. Interested readers can read anout their usage and customization at these links https://github.com/syedhali/ EZAudioandhttps://github.com/core-plot/core-plot.

3.4 REFERENCES

[1] R. Pourreza-Shahri, S. Parris, F. Saki, I. Panahi, and N. Kehtarnavaz. From Simulink to Smartphone: Signal processing application examples, *Proc. of IEEE ICASSP Conference*, Australia, April 2015. DOI: 10.1109/ICASSP.2015.7178293. 53

CHAPTER 4

Linear Time-Invariant Systems and Convolution

4.1 CONVOLUTION AND ITS NUMERICAL APPROXIMATION

The output $y(t)$ of a continuous-time linear time-invariant (LTI) system to an input $x(t)$ can be found via the convolution integral involving the input and the system impulse response $h(t)$ (for details on the theory of convolution and LTI systems, refer to signals and systems textbooks, for example, references [1–8]):

$$y(t) = \int_{-\infty}^{\infty} h(t - \tau)x(\tau)d\tau. \tag{4.1}$$

For a computer program to perform the above continuous-time convolution integral, a numerical approximation of the integral is needed noting that computer codes operate in a discrete and not continuous fashion. One way to approximate the continuous functions in the integral in Equation (4.1) is to use piecewise constant functions by defining $\delta_\Delta(t)$ to be a rectangular pulse of width Δ and height 1, centered at $t = 0$, as follows:

$$\delta_\Delta(t) = \begin{cases} 1 & -\Delta/2 \leq t \leq \Delta/2 \\ 0 & \text{otherwise.} \end{cases} \tag{4.2}$$

Then, a continuous function $x(t)$ can be approximated with a piecewise constant function $x_\Delta(t)$ as a sequence of pulses that are spaced every Δ seconds in time with heights of $x(k\Delta)$, i.e.,

$$x_\Delta(t) = \sum_{k=-\infty}^{\infty} x(k\Delta)\delta_\Delta(t - k\Delta). \tag{4.3}$$

In the limit as $\Delta \to 0$, $x_\Delta(t) \to x(t)$. As an example, Figure 4.1 shows the approximation of a decaying exponential function $x(t) = \exp\left(-\frac{t}{2}\right)$ starting from 0 by using $\Delta = 1$. Similarly, $h(t)$ can be approximated by

$$h_\Delta(t) = \sum_{k=-\infty}^{\infty} h(k\Delta)\delta_\Delta(t - k\Delta). \tag{4.4}$$

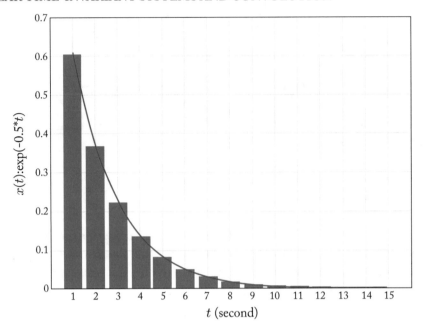

Figure 4.1: Approximation of a decaying exponential with rectangular strips of width 1.

The convolution integral can thus be approximated by convolving the two piecewise constant signals as follows:

$$y_\Delta(t) = \int_{-\infty}^{\infty} h_\Delta(t - \tau) x_\Delta(\tau) d\tau. \tag{4.5}$$

Notice that $y_\Delta(t)$ is not necessarily a piecewise constant function. For computer representation purposes, the output needs to be generated in a discrete manner. This is achieved by further approximating the convolution integral as indicated below:

$$y_\Delta(n\Delta) = \Delta \sum_{k=-\infty}^{\infty} x(k\Delta) h((n - k)\Delta). \tag{4.6}$$

By representing the signals $h_\Delta(t)$ and $x_\Delta(t)$ in a .m file as vectors containing the values of the signals at $t = n\Delta$, Equation (4.5) needs to be implemented to compute an approximation to the convolution of $x(t)$ and $h(t)$. The discrete convolution sum $\sum_{k=-\infty}^{\infty} x(k\Delta)h((n - k)\Delta)$ can be computed with the MATLAB function `conv`. Then, this sum can be multiplied by Δ to get an estimate of the continuous signal $y(t)$ at $t = n\Delta$. Note that as Δ is made smaller, one obtains a closer approximation to $y(t)$.

Note that throughout the book, the notations dt, delta, and Δ are used interchangeably to denote the time interval between samples.

Commutative

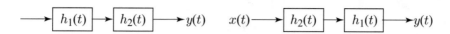

Associative

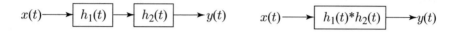

Distributive

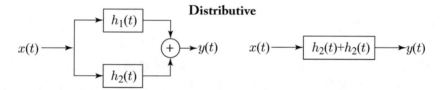

Figure 4.2: Convolution properties.

4.2   CONVOLUTION PROPERTIES

Convolution possesses the following three properties (see Figure 4.2):

Commutative property

$$x(t) * h(t) = h(t) * x(t).$$ (4.7)

Associative property

$$x(t) * h_1(t) * h_2(t) = x(t) * \{h_1(t) * h_2(t)\}.$$ (4.8)

Distributive

$$x(t) * \{h_1(t) + h_2(t)\} = x(t) * h_1(t) + x(t) * h_2(t).$$ (4.9)

4.3   CONVOLUTION EXPERIMENTS

This lab involves experimenting with the convolution of two continuous-time signals. The equation part is written as a .m file, which is then implemented and run on a smartphone platform. Due to the discrete-time nature of programming, an approximation of the convolution integral is made.

```
Editor - C:\Users\Axa180003\Desktop\MATLAB Examples\Chapter 4\L4_1\L4_1.m                    ⊙ ✕
L4_1.m  ✕  +
1     function [x1,x2,y,y_ac,MSE]=L4_1(a,b,Delta)
2
3 -   t=0:Delta:8; %Time at which samples are generated
4 -   Lt=length(t);
5
6 -   x1=exp(-a*t); %First exponential function
7 -   x2=exp(-b*t); %Second exponential function
8 -   y=Delta*conv(x1,x2); %Convolution of the two functions
9 -   y_ac=1/(a-b)*(exp(-b*t)-exp(-a*t)); %Obtaining the analytical convolution
10 -  MSE=sum((y(1:Lt)-y_ac).^2)/Lt; %Mean square error between the two convolution results
```

Figure 4.3: L4_1.m script.

L4.1 NUMERICAL APPROXIMATION OF CONVOLUTION

In this section, let us apply the MATLAB function `conv` to compute the convolution of two signals. One can choose various values of the time interval Δ to compute different or coarse to fine approximations to the convolution integral.

In this example, the function `conv` is used to compute the convolution of the signals $x(t) = \exp(-at)u(t)$ and $h(t) = \exp(-bt)u(t)$, where $u(t)$ represents a step function starting at 0 for $0 \leq t \leq 8$. Consider the following values of the approximation pulse width or delta: $\Delta = 0.5, 0.1, 0.05, 0.01, 0.005, 0.001$. Mathematically, the actual convolution of $h(t)$ and $x(t)$ is given by

$$y(t) = \frac{1}{a-b}\left(e^{-at} - e^{-bt}\right)u(t). \tag{4.10}$$

Compare the approximation $\hat{y}(n\Delta)$ obtained via the function `conv` with the actual values given by Equation (4.10). To better see the difference between the approximated $\hat{y}(n\Delta)$ and the true $y(n\Delta)$ values, it helps to display $\hat{y}(t)$ and $y(t)$ in the same graph.

Next, let us compute the mean squared error (*MSE*) between the true and approximated values using the following equation:

$$MSE = \frac{1}{N}\sum_{n=1}^{N}(y(n\Delta) - \hat{y}(n\Delta))^2, \tag{4.11}$$

where $N = \lfloor \frac{T}{\Delta} \rfloor$, T is an adjustable time duration expressed in seconds, and the symbol $\lfloor . \rfloor$ denotes the nearest integer. To begin with, let us set $T = 8$.

Open MATLAB (version 2015b or a later version), start a new MATLAB script in the HOME panel, and select the New Script. Write the following MATLAB code and save the function by using the name L4_1.m.

```
   Editor - C:\Users\Axa180003\Desktop\MATLAB Examples\Chapter 4\L4_1\L4_1_testbench.m                      ⊙ ×
   L4_1_testbench.m  ×   +
 1─    clear;
 2─    clc;
 3
 4─    Delta=0.05; %Time interval between samples
 5─    a=2; %Decay rate of first exponential function
 6─    b=3; %Decay rate of second exponential function
 7
 8─    [x1,x2,y,y_ac,MSE]=L4_1(a,b,Delta); %Calling the function which performs operations
 9
10     %% Verification of results
11─    tx = linspace(0,8,(8+Delta)/Delta);
12─    figure;
13─    plot(tx,x1,'linewidth',1);
14─    hold all
15─    plot(tx,x2,'linewidth',1)
16     %legend('x1','x2','fontsize',12)
17─    xlabel('t','fontsize',12)
18─    ty = linspace(0,16,(16+Delta)/Delta);
19─    figure;
20─    plot(ty,y,'linewidth',1);
21─    hold all
22─    plot(tx,y_ac,'linewidth',1)
23     %legend('y','y-ac','fontsize',12)
24─    xlabel('t','fontsize',12)
25─    display(MSE);
```

Figure 4.4: L4_1_testbench script.

The above code first generates a time vector t based on a time interval `Delta` for 8 s. The two input signals, `x1` and `x2`, are then convolved using the function `conv`. Next, the actual output `y_ac` based on Equation (4.10) is computed. The length of the time vector Lt and the input vectors are obtained by using the command `length(t)`. Note that the output vector `y` has a different size (for two input vectors of size m and n, the vector corresponding to the convolution output is of size $m + n - 1$). Thus, to have the same size for the output, use the same portion of the convolution output corresponding to `y(1:Lt)` for the error calculation.

Next, write a script for testing purposes. Open a *New Script*, write your code, and save it using the name L4_1_testbench, as shown in Figure 4.4.

Make sure that the function L4_1 and the script L4_1_testbench appear in the same directory. The outcome can be verified by plotting `x1`, `x2`, `y`, and `y_ac`. Run the script L4_1_testbench. Figure 4.5 shows an example outcome for the values specified in L4_1_testbench.

Next, use the MATLAB Coder to generate the corresponding C code. From the APP panel, select the MATLAB Coder and follow the steps covered in Chapter 3 to generate the corresponding C code as outlined in Figure 4.6.

Figure 4.6 shows the initial screen for the Coder. Start by selecting the function to be converted and change the Numeric Conversion option to single precision floating-point arithmetic.

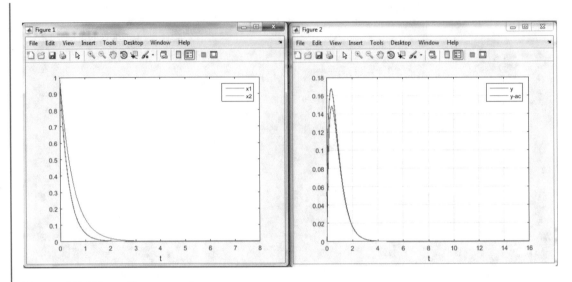

Figure 4.5: Plots of inputs and outputs.

Figure 4.6: MATLAB Coder function selection.

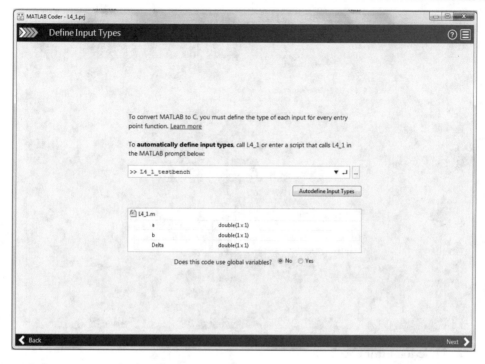

Figure 4.7: Input type specification.

After the function is selected, the input types need to be specified (see Figure 4.7). This can be done manually or by using the above test bench script to automatically determine the data types. Select the test bench script and choose the option *Autodefine Input Types* to complete this step.

After the input types are set, the Coder then checks to ensure that it is able to generate a C code from the provided MATLAB script. Figure 4.8 shows the outcome with no detected errors.

Once the MATLAB script is checked and passed, a corresponding C source code is generated by pressing the *Generate* button (see Figure 4.9). Although various configuration settings are available, the default settings are adequate for our purposes. After this step, a folder named *codegen* is created in the directory of the MATLAB files.

The final step to implement an algorithm written in MATLAB on a smartphone is to deploy the generated C code on a suitable smartphone target device. For this step, a shell is provided here in which the generated C source code needs to be placed. This shell operates in the same manner as the test bench script stated earlier; that is the two signals (a pulse and an exponential) are generated. Figures 4.10 and 4.11 show how the generated C code is integrated into the provided Android app shell. The following steps need to be taken.

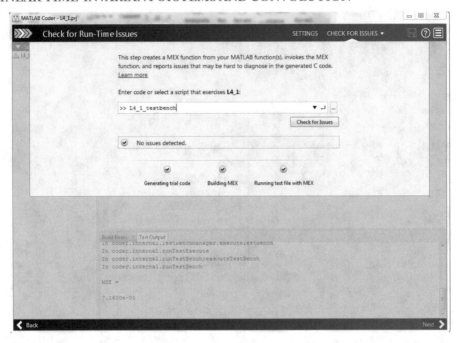

Figure 4.8: Function error check.

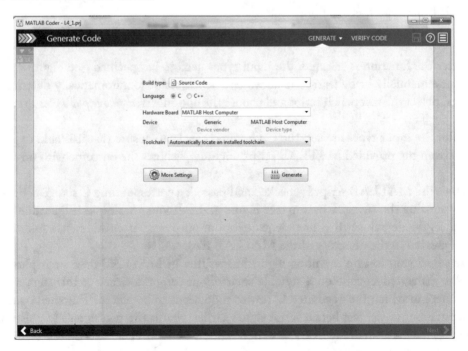

Figure 4.9: C source code generation.

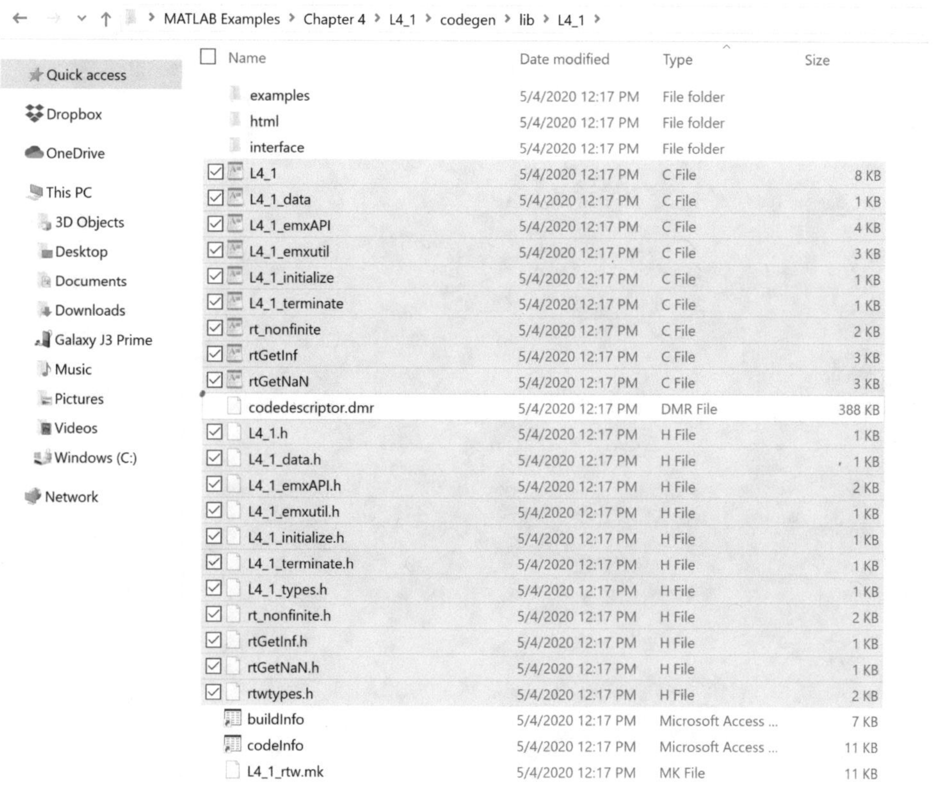

Figure 4.10: Copying files.

1. Navigate to the folder named L4_1 in the *codegen* folder; *codegen/lib/L4_1*. Copy all the files with .h and .c extensions, as illustrated in Figure 4.10.

2. If using MATLAB R2016b or later, it is required to add a specific header file into the *jni* folder separately. This header file (`tmwtypes.h`) can be found in the MATLAB root with the following path:

```
MATLAB root\R2019b\extern\include
```

3. Place the copied files inside the *jni* folder of the shell provided. The *jni* folder resides at *app/src/main/jni* (see Figure 4.11).

4. Before building the project, first press clean project (see Figure 4.12).

5. Then, run the project by pressing the Run button.

Figure 4.11: C source code integration.

Figure 4.12: **Clear project.**

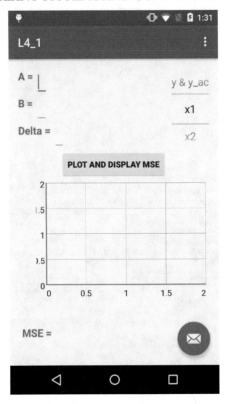

Figure 4.13: Initial screen of the app on an Android smartphone.

When you run the project, a graphical display on the connected smartphone should pop up. Figure 4.13 shows the graphical display screen for the app. Enter some desired values for a, b and Delta, and then press the button PLOT AND DISPLAY MSE. As a result, y, y_ac, and MSE for a=2, b=3, and Delta= 0.05 get plotted; see Figure 4.14.

L4.2 CONVOLUTION EXAMPLE 2

Next, consider the convolution of the two signals $x(t) = \exp(-2t)u(t)$ and $h(t) = rect\left(\frac{t-2}{2}\right)$ for $0 \le t \le 10$, where $u(t)$ denotes a step function at time 0 and rect a rectangular function defined as

$$rect(t) = \begin{cases} 1 & -0.5 \le t < 0.5 \\ 0 & \text{otherwise.} \end{cases} \tag{4.12}$$

Let $\Delta = 0.01$. Figures 4.15 and 4.16 show the MATLAB function and the test bench script for this example. Open a *New Script* and place the code in Figure 4.15, then save the

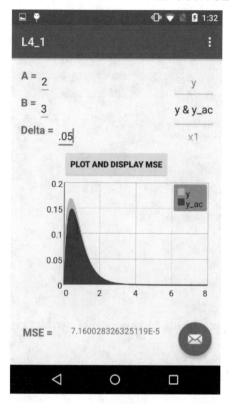

Figure 4.14: Plots of y and y_ac.

function using the name L4_2. Open another *New Script* and place the code in Figure 4.16 as the test bench code, and use the name L4_2_testbench to save this script.

Follow the steps as outlined above for L4_1 to generate the corresponding C code and then place it into the shell provided. Figures 4.17 and 4.18 show the initial screen of the app on an Android smartphone as well as $x(t)$, $h(t)$, and $x(t) * h(t)$ for `Delta =0.01` in different graphs.

L4.3 CONVOLUTION EXAMPLE 3

In this third example, the convolution of the signals shown in Figure 4.19 is performed.

Figures 4.20 and 4.21 show the MATLAB function and the test bench script for this example. Open a *New Script* and place the code listed in Figure 4.20, then save the function using the name L4_3. Open another *New Script* and place the code listed in Figure 4.21 as the test bench code, and use the name L4_3_testbench to save this script.

```
Editor - C:\Users\Axa180003\Desktop\MATLAB Examples\Chapter 4\L4_2\L4_2.m                    ⊙ ×
 L4_2.m  ×  +
 1     □ function [x,h,con]=L4_2(Delta)
 2
 3 -     t=0:Delta:10; %Time at which samples are generated
 4 -     x=exp(-2*t); %Generating the input signal
 5
 6       % Generating the impulse response
 7 -     h1=zeros(1,round(1/Delta));
 8 -     h2=ones(1,round(2/Delta));
 9 -     h3=zeros(1,round(7/Delta));
10 -     h=[h1 h2 h3];
11
12       % Obtaining the output using convolution
13 -     con=Delta*conv(x,h);
```

Figure 4.15: L4_2 function performing convolution of two signals.

```
Editor - C:\Users\Axa180003\Desktop\MATLAB Examples\Chapter 4\L4_2\L4_2_testbench.m          ⊙ ×
 L4_2_testbench.m  ×  +
 1 -   clear;
 2 -   clc;
 3
 4 -   Delta=0.01;            %Time interval between samples
 5
 6 -   [x,h,y]=L4_2(Delta); %Calling the function which performs operations
 7
 8     %% Verification of results
 9 -   figure(1)
10 -   plot(Delta*(1:length(x)),x);grid on;
11 -   figure(2)
12 -   plot(Delta*(1:length(h)),h);grid on;
13 -   figure(3)
14 -   plot(Delta*(1:length(y)),y);grid on;
```

Figure 4.16: L4_2_testbench.

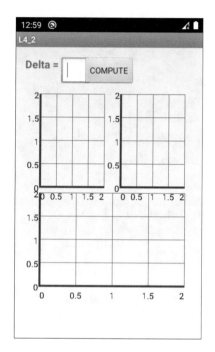

Figure 4.17: Initial screen of the app on an Android smartphone.

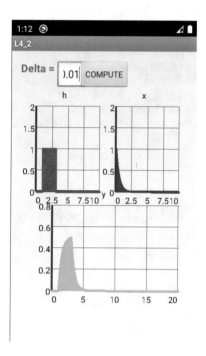

Figure 4.18: Plots of $x(t)$, $h(t)$, and $y(t)$.

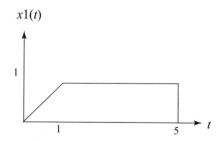

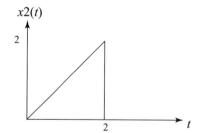

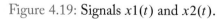

Figure 4.19: Signals $x1(t)$ and $x2(t)$.

Editor - C:\Users\Axa180003\Desktop\MATLAB Examples\Chapter 4\L4_3\L4_3.m

L4_3.m

```
1    function [x1,x2,y]=L4_3(dt)
2
3    % Generating the first signal
4    a=0:dt:1;
5    b=ones(1,round(4/dt));
6    x1=[a b];
7
8    % Generating the second signal
9    x2=0:dt:2;
10
11   % Convoluting
12   y=dt*conv(x1,x2);
```

Figure 4.20: L4_3 function performing convolution of two signals.

Editor - C:\Users\Axa180003\Desktop\MATLAB Examples\Chapter 4\L4_3\L4_3_testbench.m

L4_3_testbench.m

```
1    clear;
2    clc;
3
4    Delta=0.001;          %Time interval between samples
5
6    [x1,x2,y]=L4_3(Delta); %Calling the function which performs operations
7
8    %% Verification
9    figure(1)
10   plot(Delta*(1:length(x1)),x1);grid on;
11   figure(2)
12   plot(Delta*(1:length(x2)),x2);grid on;
13   figure(3)
14   plot(Delta*(1:length(y)),y);grid on;
```

Figure 4.21: L4_3_testbench.

Figure 4.22 shows the smartphone screen for the third convolution example and Figure 4.23 shows the plots of the signals $x1(t)$, $x2(t)$, and $x1(t) * x2(t)$ in separate graphs.

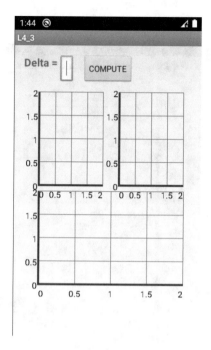

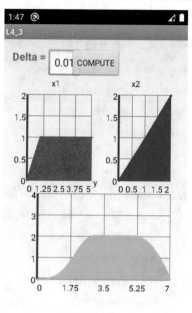

Figure 4.22: Initial screen of the app on an Android smartphone.

Figure 4.23: Plots of $x(t)$, $h(t)$, and $y(t)$.

L4.4 CONVOLUTION PROPERTIES

In this section, the convolution properties are examined. Figures 4.24 and 4.25 show the MATLAB function and the test bench script for this example. Open a *New Script* and place the code listed in Figure 4.24, then save the function using the name L4_4. Open another *New Script* and place the code listed in Figure 4.25 as the test bench code, and use the name L4_4_testbench to save this script.

Both sides of Equations (4.7), (4.8), and (4.9) are implemented in the function L4_4, where the commutative, associative, and distributive properties of convolution are illustrated, respectively. In this code, first the signals $x(t)$, $h1(t)$, and $h2(t)$ are generated. Then, the output signal $y(t)$ is generated depending on the user choice of the properties. Here, the MATLAB command `switch` is used. This command evaluates an expression, and chooses to execute one set of statements out of several sets of statements.

```
Editor - C:\Users\Axa180003\Desktop\MATLAB Examples\Chapter 4\L4_4\L4_4.m
L4_4.m  ×  +
1    function [x,h1,h2,y]=L4_4(dt,choice)
2
3        % Generating the input signal
4  -     x=[ones(1,round(2/dt)) -ones(1,round(2/dt))];
5        % Generating first impulse response
6  -     h1=[0:dt:2-dt 2:-dt:dt];
7        % Generating second impulse response
8  -     h2=0:dt:4-dt;
9        % Initializing y
10 -     y=x;
11
12       % Output the correct signal based on the choice
13 -     switch choice
14 -         case 0
15 -             y=dt*conv(dt*conv(h1,x),h2);
16 -         case 1
17 -             y=dt*conv(dt*conv(h2,x),h1);
18 -         case 2
19 -             y=dt*conv(dt*conv(x,h1),h2);
20 -         case 3
21 -             y=dt*conv(dt*conv(h1,h2),x);
22 -         case 4
23 -             y=dt*conv(x,h1+h2);
24 -         case 5
25 -             y=dt*conv(x,h1)+dt*conv(x,h2);
26 -     end
```

Figure 4.24: L4_4 function incorporating convolution properties.

```
Editor - C:\Users\Axa180003\Desktop\MATLAB Examples\Chapter 4\L4_4\L4_4_testbench.m
L4_4_testbench.m  ×  +
1  -   clear;
2  -   clc;
3
4  -   Delta=0.001; %Time interval between samples
5  -   choice=2; % choice for the property to check
6      % 0:(h1*x)*h2, 1:(h2*x)*h1, 2:(x*h1)*h2, 3:(h1*h2)*x, 4:x*(h1+h2), 5:(x*h1)+(x*h2)
7
8  -   [x,h1,h2,y] = L4_4(Delta,choice); %Calling the function which performs operations
9
10     %% Verification of results
11 -   figure(1)
12 -   plot(Delta*(1:length(x)),x);grid on;
13 -   figure(2)
14 -   plot(Delta*(1:length(h1)),h1);grid on;
15 -   figure(3)
16 -   plot(Delta*(1:length(h2)),h2);grid on;
17 -   figure(4)
18 -   plot(Delta*(1:length(y)),y);grid on;
```

Figure 4.25: L4_4_testbench for plotting $x(t)$, $h1(t)$, $h2(t)$, and $y(t)$ for choice 2.

```
switch switch_expression
    case case_expression
        statements
    case case_expression
        statements
    ...
    otherwise
        statements
end
```

In the function L4_4, `switch_expression` provides choices. A choice is denoted by an integer number. As the function L4_4 is called for a specific choice value, MATLAB tests each case until one of `case expressions` is true, that is `case_expression = choice value`, and then the corresponding statements for that `case_expression` are executed.

Note any variable in a MATLAB function that is not directly assigned by a MATLAB command needs to be initialized; otherwise the MATLAB Coder will throw an error that the C code cannot be generated. For the examples mentioned here, the variable $y(t)$ is not directly assigned in a MATLAB command and its value is determined in a switch case statement. Therefore, it is necessary to first initialize it to prevent such an error. This variable is initialized to be $x(t)$ here.

After running L4_4_testbench, follow the steps in L4_1 to generate the corresponding C code and then place it into the shell provided. Figures 4.26–4.35 show the app screen on an Android smartphone as well as $x(t)$, $h(t)$, and $y(t)$ plots for different *Delta*. A desired plot can be chosen from a choice picker located on the right side and by pressing the PLOT button.

L4.5 LINEAR CIRCUIT ANALYSIS USING CONVOLUTION

In this section, let us consider an application of convolution involving RLC linear circuits to gain a better understanding of the convolution concept. A linear circuit is an example of a linear system, which is characterized by its impulse response $h(t)$, that is, the output in response to a unit impulse input. The input to such circuits can be considered to be an input voltage $v(t)$ and the output to be the output current $i(t)$, as illustrated in Figure 4.36.

For a simple RC series circuit shown in Figure 4.37, the impulse response is given by [8]:

$$h(t) = \frac{1}{R} \exp\left(-\frac{1}{RC}t\right), \tag{4.13}$$

which can be obtained for any specified values of R and C. When an input voltage $v(t)$ (either DC or AC) is applied to the circuit, the current $i(t)$ can be obtained by simply convolving the

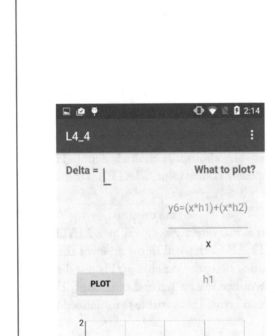

Figure 4.26: L4_4 initial screen.

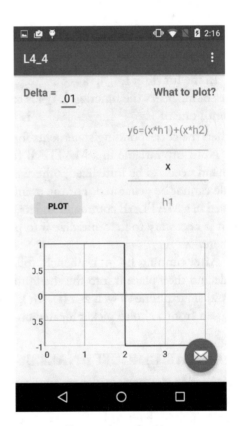

Figure 4.27: L4_4 plot of $x(t)$.

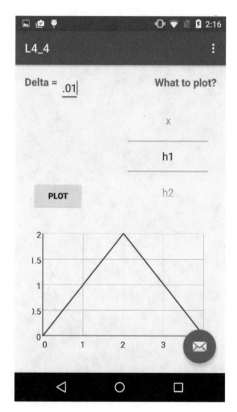

Figure 4.28: Plot of $h1(t)$.

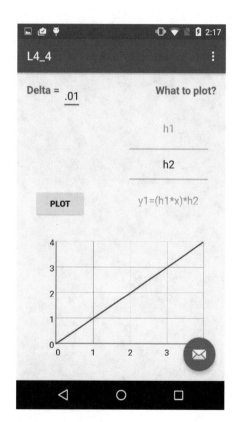

Figure 4.29: Plot of $h2(t)$.

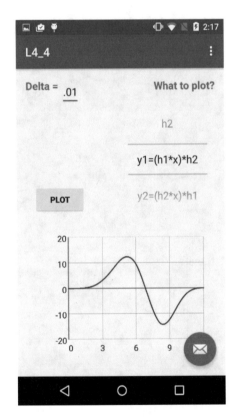

Figure 4.30: Commutative property.

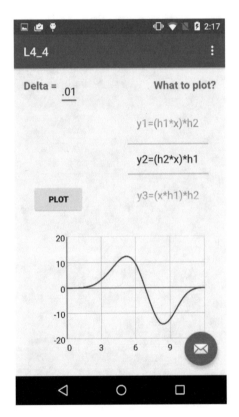

Figure 4.31: Commutative property.

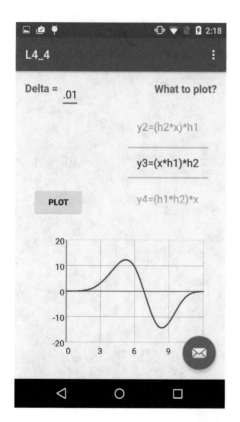

Figure 4.32: **Associative** property.

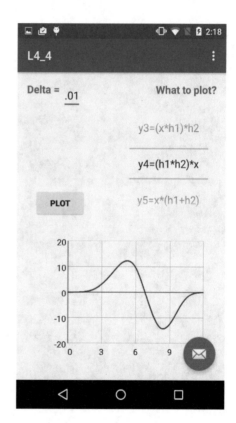

Figure 4.33: **Associative** property.

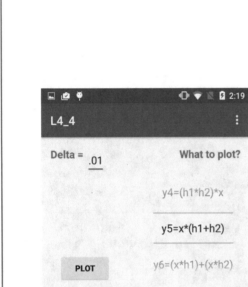

Figure 4.34: Distributive property.

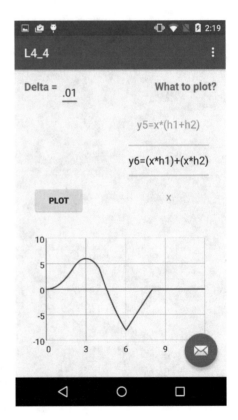

Figure 4.35: Distribute property.

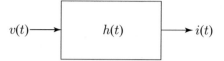

Figure 4.36: Impulse response representation of a linear circuit.

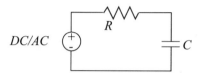

Figure 4.37: RC circuit.

circuit impulse response with the input voltage, that is

$$i(t) = h(t) * v(t). \tag{4.14}$$

Similarly, for a simple RL series circuit shown in Figure 4.38, the impulse response is given by

$$h(t) = \frac{1}{L} \exp\left(-\frac{R}{L}t\right). \tag{4.15}$$

When an input voltage $v(t)$ is applied to the circuit, the current $i(t)$ can be obtained by computing the convolution integral.

Figures 4.39 and 4.40 show the MATLAB function and the test bench script for this example. Open a *New Script* and place the code listed in Figure 4.39, then save the function using the name L4_5. Open another *New Script* and place the code listed in Figure 4.40 as the test bench code, and use the name L4_5_testbench to save this script. The function L4_5 has the inputs: `dt`, `w1`, `w2`, `A`, `R`, `L`, and `C`, where `dt` stands for the time interval Δ, `w1`, and `w2` denote the input voltage type (DC, `w1=0` or `w1= 1`, AC) and the circuit type (RC, `w2=0` or `w2= 1`, RL), respectively. The notations `A`, `R`, `L`, and `C` denote voltage amplitude, resistance, inductance, and capacitance values, respectively. Make sure that the order of the inputs is kept as it appears in the MATLAB code.

After running L4_5_testbench, follow the steps outlined in L4_1 to generate the C code and then place it into the shell provided. Figures 4.41–4.46 show the app screen on an Android smartphone as well as $x(t)$, $h(t)$, and $y(t)$ for *Delta* $= 0.01$.

From the app, one can control the circuit type (RL or RC), the input voltage type (DC or AC) and the input voltage amplitude. One can also observe the circuit response by changing R, L, and C values. After entering the desired settings, the desired *Delta* and *Amplitude* of the input signal and the desired R, C, and L values, press the COMPUTE button.

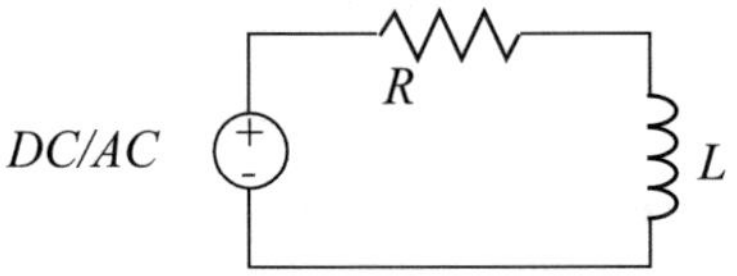

Figure 4.38: **RL** circuit.

```
Editor - C:\Users\Axa180003\Desktop\MATLAB Examples\Chapter 4\L4_5\L4_5.m                          ⊙ ✕
L4_5.m  ✕  +
1     function [x,h,y]=L4_5(dt,w1,w2,A,R,L,C)
2
3 -   t=0:dt:10;
4
5     % Generating the input signal
6 -   if w1==0
7 -   x=A*ones(1,length(t));
8 -   else
9 -   x=A*sin(4*t);
10 -  end
11
12 -  if w2==0
13    % RL response
14 -  h=R/L*exp(-R/L*t); % Impulse response corresponding to RL system
15 -  y=dt*conv(x,h); % Output
16 -  else
17    % RC response
18 -  h=1/(R*C)*exp(-t/(R*C)); % Impulse response corresponding to RC system
19 -  y=1/(R*C)-dt*conv(x,h); %Output
20 -  end
21 -  y=y(1:round(10/dt));
```

Figure 4.39: **L4_5** function for the linear circuit.

Figure 4.40: **L4_5_testbench** for plotting $x(t)$, $h(t)$, and $y(t)$ of RC circuit.

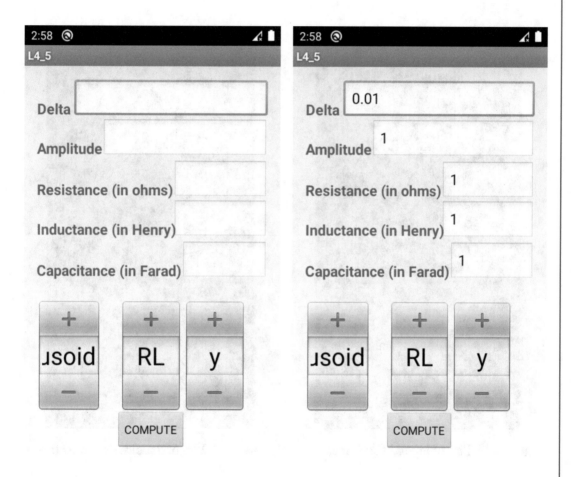

Figure 4.41: Smartphone app screen.   Figure 4.42: Settings screen.

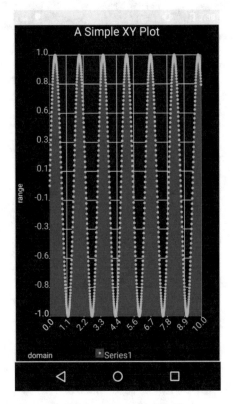

Figure 4.43: Plot of input AC voltage.

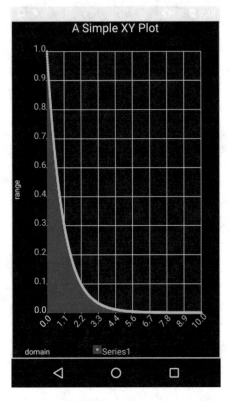

Figure 4.44: Plot of impulse response $h(t)$ for RL circuit.

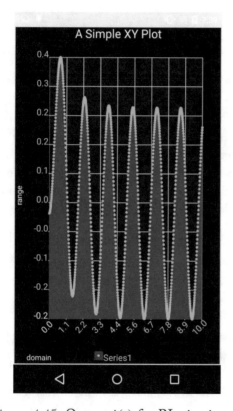

Figure 4.45: Output $i(t)$ for RL circuit and AC input voltage.

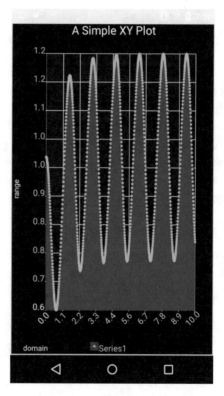

Figure 4.46: Output $i(t)$ for RC circuit and AC input voltage.

4.4   LAB EXERCISES

4.4.1   ECHO CANCELLATION

In this exercise, the problem of removing an echo from a recording of a speech signal is considered. The MATLAB function `sound()` can be used to play back a speech recording. Load the file *echo_1.wav* provided in the book software package by using the MATLAB function `audioread('filename')`. This speech file was recorded at the sampling rate of 8 kHz, which can be played back through the computer speakers by typing

```
>> sound(y)
```

You should be able to hear the sound with an echo. Build a waveform based on the loaded data and the time interval $dt = 1/8000$ noting that this speech was recorded using an 8 kHz sampling rate.

An echo is produced when the signal (speech, in this case) is reflected off a non-absorbing surface like a wall. What is heard is the original signal superimposed on the signal reflected off the wall (echo). Because the speech is partially absorbed by the wall, it decreases in amplitude. It is also delayed. The echoed signal can be modeled as $ax(t - \tau)$, where $a < 1$ and τ denotes the echo delay. Thus, one can represent the speech signal plus the echoed signal as [2]

$$y(t) = x(t) + ax(t - \tau). \tag{4.16}$$

What is heard is $y(t)$. It is desired here to recover $x(t)$—the original, echo-free signal—from $y(t)$.

Method 1

In this method, remove the echo using deconvolution. Rewrite Equation (4.16) as follows [2]:

$$y[n\Delta] = x[n\Delta] + ax[(n - N)\Delta] = x[n\Delta] * (\delta[n\Delta] + a\delta[n - N]\Delta) = x[n\Delta] * h[n\Delta]. \tag{4.17}$$

The echoed signal is the convolution of the original signal $x(n\Delta)$ and the signal $h(n\Delta)$. Use the function `deconv(y,h)` to recover the original signal.

Method 2

An alternative way of removing the echo is to run the echoed signal through the following linear system:

$$z[n\Delta] = y[n\Delta] - az[(n - N)\Delta]. \tag{4.18}$$

Assume that $z[n\Delta] = 0$ for $n < 0$. Implement the above system for different values of a and N.

Display and play back the echoed signal and the echo-free signal using both of the above methods. Specify the parameters a and N as control parameters. Try to measure the proper values of a and N by the autocorrelation method described below.

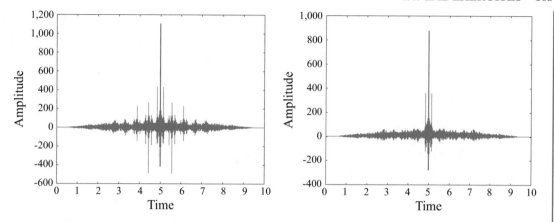

Figure 4.47: Autocorrelation function of a signal: (left figure) echo is partially removed; (right figure) echo is mostly removed.

The autocorrelation of a signal can be described by the convolution of a signal with its mirror. That is,

$$R_{xx}[n] = x[n] * x[-n]. \qquad (4.19)$$

Use the autocorrelation of the output signal (echo-free signal) to estimate the delay time N and the amplitude of the echo a. For different values of N and a, observe the autocorrelation output. To have an echo-free signal, the side lobes of the autocorrelation should be quite low, as indicated in Figure 4.47.

4.4.2 NOISE REDUCTION USING MEAN FILTERING

The idea of mean filtering is simply to replace each value in a signal with the mean (average) value of its neighbors. A mean filter is widely used for noise reduction.

Start by adding some random noise to a signal (use the file *echo_1.wav* or any other speech data file). Then, use mean filtering to reduce the introduced noise. More specifically, take the following steps.

1. Normalize the signal values in the range [0 1].

2. Add random noise to the signal by using the function `randn`. Set the noise level as a control parameter.

3. Convolve the noise-added signal with a mean filter. This filter can be designed by taking an odd number of ones and dividing the sum by the size. For example, a 1×3 size mean filter is given by [1/3 1/3 1/3] and a 1×5 size mean filter by [1/5 1/5 1/5 1/5 1/5]. Set the size of the mean filter as an odd number control parameter (3, 5, or 7, for example).

4.4.3   IMPULSE NOISE REDUCTION USING MEDIAN FILTERING

A median filter is a nonlinear filter that replaces a data value with the median of the values within a window. For example, the median value for this data stream [2 5 3 11 4] is 4. This type of filter is often used to remove impulse noise. Use the file *echo_1.wav* or any other speech data file and take the following steps.

1. Normalize the signal values in the range [0 1].

2. Randomly add impulse noise to the signal by using the MATLAB function `randperm`. Set the noise density as a control parameter.

3. Find the median values using the function *median* and replace the original value with the median value. Set the number of the window as an odd number control parameter (3, 5, or 7, for example).

4.5   RUNNING MATLAB CODER-GENERATED C CODES ON SMARTPHONES

4.5.1   RUNNING ON ANDROID SMARTPHONES

This section covers the steps for integrating a C code generated by the MATLAB Coder into an Android shell program for running it on an Android smartphone. These steps are listed as follows.

1. Creating a shell.

2. Verification of the MATLAB function to be run on smartphones.

3. Using MATLAB Coder to generate the corresponding C code to be placed in the shell.

4. Modifications of the shell to integrate the C code into it.

The first three steps were covered earlier. In this section, the fourth step is covered. Let us consider the basic HelloWorld shell that was mentioned and the C code generated by the MATLAB Coder. The following modifications of the shell are needed for running the code on an Android smartphone. The HelloWorld shell for L4_1:

1. Navigate to the folder created by the MATLAB Coder named *codegen\lib\L4_1*. Copy all the files with the extensions .h and .c.

2. If using MATLAB R2016b or later, it is required to add a specific header file into the *jni* folder separately. This header file (`tmwtypes.h`) can be found in the MATLAB root with the following path:

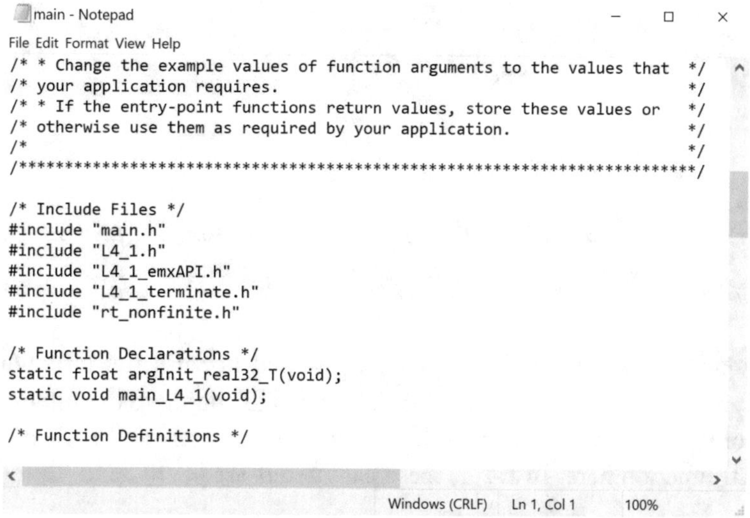

```
main - Notepad                                                    −   □   ×
File Edit Format View Help
/* * Change the example values of function arguments to the values that  */  ^
/* your application requires.                                            */
/* * If the entry-point functions return values, store these values or   */
/* otherwise use them as required by your application.                   */
/*                                                                       */
/***********************************************************************/

/* Include Files */
#include "main.h"
#include "L4_1.h"
#include "L4_1_emxAPI.h"
#include "L4_1_terminate.h"
#include "rt_nonfinite.h"

/* Function Declarations */
static float argInit_real32_T(void);
static void main_L4_1(void);

/* Function Definitions */
```
Windows (CRLF)   Ln 1, Col 1   100%

Figure 4.48: main.c file located in the folder *codegen\lib\L4_1\examples*.

```
MATLAB root\R2019b\extern\include
```

3. Place the copied files inside the *jni* folder of the HelloWorld shell.

4. Open the HelloWorld project in Android Studio.

5. Double click on the file HelloWorld.c in the *jni* folder.

   Inside the code, the following statement can be seen:

```
#import <jni.h>
jstring Java_com_dsp_helloworld_MainActivity_getString
( JNIEnv* env, jobject thiz ) {
      return (*env) -> NewStringUTF(env, "Hello UTD!");
}
```

6. Navigate to the folder *codegen\lib\L4_1\examples* and open the file main.c using Notepad.

7. Copy the part on the top (below Include Files in Figure 4.48, but not main.h) into the HelloWorld.c file of the shell.

8. Next, modify the statement line from

```
jstring Java\_com\_dsp\_helloworld\_MainActivity\_getString
    ( JNIEnv* env, jobject thiz )
```

to

```
jdoubleArray Java_com_dsp_Helloworld_MainActivity_getString
( JNIEnv* env, jobject thiz, jdouble a, jdouble b, jdouble delta )
```

Note jdoubleArray instead of jstring is used because the intention in the HelloWorld project was to return the string "Hello UTD" but here the intention is to return an array of double type numbers. Add the input parameters and their data types of the MATLAB function here. In L4_1, these parameters are a, b, and *delta* as:

```
jdoubleArray Java_com_dsp_helloworld_MainActivity_getString
( JNIEnv* env, jobject thiz, jdouble a, jdouble b, jdouble delta )
```

9. Then, inside the function created, place the following statements:

```
int length=round(16/delta);
jdoubleArray jArray = (*env) -> NewDoubleArray(env, length);
```

These statements define the output variable and allocate memory to it.

```
From the file main.c, copy the following lines that define and
initialize the variables:

    /* Initialize function 'L4_1' input arguments.  */
    emxArray_real32_T *x1;
    emxArray_real32_T *x2;
    emxArray_real32_T *y;
    emxArray_real32_T *y_ac;
    float MSE;
    emxInitArray_real32_T(&x1, 2);
    emxInitArray_real32_T(&x2, 2);
    emxInitArray_real32_T(&y, 2);
```

```
emxInitArray_real32_T(&y_ac, 2);
```

Now copy the following line from the file main.c and place it in the HelloWorld file:

```
/* Call the entry-point 'L4_1'. */
L4_1(a, b, delta, x1, x2, y, y_ac, &MSE);
```

This function calls the function L4_1, where the five output arrays `x1`, `x2`, `y`, `y_ac`, and `MSE` are returned. To assign y data values to *Array*, a temporary array called `y_temp` is used in a for loop and is filled by y data values:

```
int i;
doubley_temp[length];
for( i = 0; i < length; i++) {
    y_temp[i] = y -> data[i];
    // __android_log_print(ANDROID_LOG_ERROR,
    //    "HelloWorld Final 1", "%f\n", y\_temp[i]);
}
```

Note that for `MSE`, this step is not needed since it is only one value.

Next, free the allocated memory by copying the following lines from the file main.c:

```
emxDestroyArray_real32_T(y_ac);
emxDestroyArray_real32_T(y);
emxDestroyArray_real32_T(x2);
emxDestroyArray_real32_T(x1);
```

Assign `y_temp` to `Array` and return it:

```
    (*env) -> SetDoubleArrayRegion(env,jArray,0,length,y\_temp);
    return jArray;
}
```

These steps need to be repeated for all the defined outputs in the function L4_1; see the shell provided for L4_1.

10. Finally, design your desired Graphical User Interface (GUI). This link provides guidelines as how to design GUIs: http://androidplot.com/docs/quickstart/

4.5.2   RUNNING ON iOS SMARTPHONES

This section covers the steps for integrating a C code generated by the MATLAB Coder into the Xcode shell previously covered for running it on iPhone smartphones. These steps are listed below.

1. Create the shell.

2. Verify the MATLAB function to be run.

3. Use the MATLAB Coder to generate the corresponding C code to be placed in the shell.

4. Modify the shell to integrate the C code into it.

The first three steps were covered previously in the Android section and are the same for iOS. In this section, the fourth step is covered. Let us consider the lab L4_1 shell and its corresponding C code generated by the MATLAB Coder. The following modifications of the shell are needed for running the code on an iOS smartphone.

Here it is worth mentioning that the group structure in Xcode is not the same as the folder structure in Finder on Mac. It can be made to be the same, but making a group does not make a relative folder on the file system.

5. Make the directory *codegen/lib/L4_1* as a subfolder in the Xcode project L4_1.

6. Navigate to the folder created by the MATLAB Coder named *codegen\lib\L4_1*. Copy all the files with the extensions .h and .c to the Xcode project.

7. In the Xcode Project Navigator, click on *File -> Add Files to "L4_1"*.

8. Click *File -> New -> Group* and type in "Native Code".

9. Select all of the .c and .h files and drag them into the *"Native Code"* group.

10. If using MATLAB R2016b or later, it is required to add a specific header file into the *"Native Code"* group separately. This header file (`tmwtypes.h`) can be found in the MATLAB root with the following path:

```
MATLAB root\R2019b\extern\include
```

The MATLAB C codes will be built and get ready for use. The code to interact with buttons and graphs was covered earlier. Now, the generated C code needs to get placed or integrated into the iOS shell.

Click on "ViewController.m" to edit the file. At the top of the file above the interface definition, import the following necessary header files:

```
#import "ViewController.h"
#import "L4_1.h"
#import "L4_1_emxAPI.h"
```

Underneath the line, in the buttonPress method,

```
int length = round( 8.0 / delta);
```

add the following lines:

```
if( _x1 == NULL )
{
  _x1 = emxCreate_real32_T(1,length);
}
if( _x2 == NULL )
{
  _x2 = emxCreate_real32_T(1,length);
}
if( _y == NULL )
{
  _y = emxCreate_real32_T(1,length);
}
if( _y_ac == NULL )
{
  _y_ac = emxCreate_real32_T(1,length);
}

_yMax = 0;
L4_1(a, b, delta, _x1, _x2, _y, _y_ac, &mse);

[_mseLabel setText:[NSString stringWithFormat:@"%f", _mse ]];
[self.graph1 reloadData];
```

The `"if block"` lines allocate memory for the output `x1`, `x2`, `y`, and `y_ac`. The Lab 4 MATLAB code will populate these variables as well as `MSE` variable mse. Since the variables are made as member variables to ViewController, they become accessible throughout with the underscore prefix, e.g., `_x1` . This allows selecting these variables to be plotted at a later time. Calling reloadData on the graph1 object calls the delegate method "numberForPlot" which sets the appropriate domain and range for plotting.

4.6   REAL-TIME RUNNING ON SMARTPHONES

A distinguishing attribute of the introduced smartphone-based signals and systems laboratory paradigm from a conventional signals and systems laboratory paradigm is the ability to run signals and systems code written in MATLAB in real time on smartphones. This section presents the steps one needs to take in order to process audio signals captured by a smartphone microphone or a saved wave file in real-time through a linear time-invariant system implemented in MATLAB. In what follows, it is explained how to integrate a C code generated by the MATLAB Coder into a real-time shell allowing its real-time execution on Android and iOS smartphones.

4.6.1   MATLAB FUNCTION DESIGN

Let us consider the convolution algorithm written as a MATLAB script. The first step is to open MATLAB and create a new function file to run the algorithm on a frame by frame basis, that is by examining one frame of data samples at a time. The code appearing in Figure 4.49 is a frame-based implementation. Of particular importance in this function is the usage of the persistent variable `buffer` . This variable stores previous samples of an input signal between calls to the `convolution` function in order to produce the output.

4.6.2   TEST BENCH

As noted earlier, for debugging purposes as well as simulating the response of a linear time-invariant system on a target platform, a test bench MATLAB script needs to be written. For audio signal processing, a typical MATLAB script assumes samples of an entire audio signal are available but on an actual smartphone target, audio signal processing takes place one frame at a time. Thus, one needs to modify a MATLAB script for frame-based processing. The following script in Figure 4.50 shows how such an implementation is achieved.

This script generates a test signal and writes it to a file on a target platform. The signal is reshaped into a matrix of frame-sized columns and transposed to form frame-sized rows as would be the behavior on a target platform. The rows are then passed to the MATLAB function for processing. Run this test bench script and follow the steps provided in Section L4.1 to generate the corresponding C code. After generating the C code, integrate it into the real-time shell that is provided. Figure 4.51 exhibits the initial screen and the setting menu of this app.

```
Editor - C:\Users\Axa180003\Desktop\MATLAB Examples\Chapter 4\LR4_2\LR4_2.m                    ⊙ ⊗ ✕
LR4_2.m  ✕  +
1      function con=LR4_2(x,Fs)
2
3 -    Delta = 1/Fs; %Sampling time
4
5      % The impulse response is designed in such that it preserves the
6      % shape given in the manual but the number of samples is fixed to 32 so
7      % that it does not violate the real time requirements.
8 -    D = floor(32/10);
9 -    h1 = zeros(1,D);
10-    h2 = ones(1,2*D);
11-    h3 = zeros(1,32-3*D);
12-    h = [h1 h2 h3];
13
14     % Obtaining frame size from the input
15-    frameSize=length(x);
16
17     % Buffer to store previous and current samples of the input
18-    persistent buffer;
19
20-    if isempty(buffer)
21-        buffer = zeros(1,frameSize+length(h));
22-    end
23-    buffer = [buffer(:,end-length(h)+1:end) x];
24
25     % Performing convolution
26-    con=Delta*conv(buffer,h);
27-    con = con(length(h)+1:length(h)+frameSize);
```

Figure 4.49: Real-time convolution MATLAB function (named LR4-2 real-time example).

```
Editor - C:\Users\Axa180003\Desktop\MATLAB Examples\Chapter 4\LR4_2\LR4_2_testbench.m          ⊙ ⊗ ✕
LR4_2_testbench.m  ✕  +
1 -    clear;
2 -    clc;
3 -    sampleRate = 8000; %Define the sample rate
4 -    sampleTime = 1/sampleRate; %Sample time is reciprocal of sample rate
5 -    frameSize = 256; %Define the frame size
6
7      % Generating the signal
8 -    time=(0:sampleTime:10*sampleRate*sampleTime);
9 -    length = floor(size(time,2)/frameSize)*frameSize;
10-    time = time(1:length);
11-    signal = 1/2*square(2*pi*600*time,50);
12-    audiowrite('Square_duty-50_600Hz_10s.wav',signal,sampleRate);
13
14     % Simulate frame-based processing
15     % Reshaping the signal for frame processing
16-    signal = reshape(signal,frameSize,[]); signal = signal';
17
18-    result = zeros(size(signal)); %Initializing
19-    for i=1:size(signal,1)
20         %Calling the function which performs convolution
21-        result(i,:)=LR4_2(signal(i,:),sampleRate);
22-    end
23     % This is just a dummy variable used to make sure
24     % that MATLAB does not hard
25     % code the variables based on the input size in testbench
26-    test=LR4_2(1:1024,sampleRate);
```

Figure 4.50: Real-time convolution testbench script. (*Continues.*)

```
Editor - C:\Users\Axa180003\Desktop\MATLAB Examples\Chapter 4\LR4_2\LR4_2_testbench.m
  LR4_2_testbench.m  ×  +
27      %% Plotting the results for verification
28 -    result = result';
29 -    result = result(:);
30 -    signal = signal';
31 -    signal = signal(:);
32
33 -    D = floor(32/10);
34 -    h1 = zeros(1,D);
35 -    h2 = ones(1,2*D);
36 -    h3 = zeros(1,32-3*D);
37 -    h = [h1 h2 h3];
38
39      % Checking the results of frame processing
40      % result=sampleTime*conv(signal,h);
41
42 -    figure;
43 -    subplot(1,3,1);
44 -    plot(signal);
45 -    plot(signal(frameSize+1:2*frameSize));
46 -    title('Input Signal');
47 -    subplot(1,3,2);
48 -    plot(h);
49 -    title('Impulse Response');
50 -    subplot(1,3,3);
51      % plot(result);
52 -    plot(result(frameSize+1:2*frameSize));
53 -    title('Convoluted Signal');
```

Figure 4.50: (*Continued.*) Real-time convolution testbench script.

- **"Output to Playback"** – Defines which signal to be played back; Original (input signal) or Filtered (output).

- **"Sampling Frequency"** – Input sampling frequency (Fs), which can be varied from 8000–48,000 Hz.

- **"Frame Size"** – Size of frame to be processed.

- **"Debugging Level"** – Defines the debugging mode; for verification purposes choose txt file to obtain a txt file of the output array.

- **"Read File Button"** – Provides sound files in .wav format that can be used as input signals; these wav files need to be stored in the filter folder of the app on the smart-phone. After selecting a sound wav file, the processing takes place and the outcome gets saved in a txt file in the filter folder (provided that the Debugging Level is selected as txt file).

- **"Start Button"** – Starts capturing sound signals from the smartphone microphone for processing.

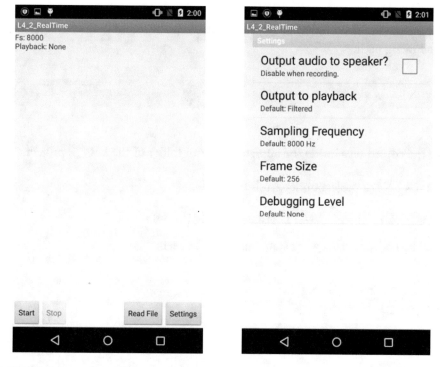

Figure 4.51: Initial screen and setting menu of real-time convolution app.

4.6.3   MODIFYING REAL-TIME SHELL FOR ANDROID

This section covers the steps for integrating a C code generated by the MATLAB Coder into the Android real-time shell program. These steps are listed below.

1. Verification of the MATLAB function to be run on an Android smartphone.

2. Using the MATLAB Coder to generate the corresponding C code.

3. Creating a real-time shell.

4. Modifications of the real-time shell to integrate the C code into it.

The first two steps were covered earlier. A generic real-time shell is provided as part of the book software package for the third step. The fourth step is covered here which involves modifications of the real-time shell for the integration of the C code. The following modifications of the shell are needed for running the C code in real-time on an Android smartphone.

1. Navigate to the folder created by the MATLAB Coder named *codegen\lib\LR4_2*. Copy all the files with the extensions .h and .c.

2. If using MATLAB R2016b or later, it is required to add a specific header file into the *jni* folder separately. This header file (`tmwtypes.h`) can be found in the MATLAB root with the following path:

```
MATLAB root\R2019b\extern\include
```

3. Place the copied files inside the app/src/main/jni folder of the LR4_2 project.

4. Open the *LR4_2* project under Android Studio.

5. Double click on the file MATLABNavtive.c in the *jni* folder.

Inside the code, the following statements can be seen:

```c
#include <jni.h>
#include <stdio.h>

jfloatArray
Java_com_dsp_matlab_Filters_compute(JNIEnv *env, jobject thiz,
jfloatArray input)
{
    jfloatArray output = (*env)->NewFloatArray(env, 256);
    float *_output =
                (*env)->GetFloatArrayElements(env, output, NULL);
    float *_in = (*env)->GetFloatArrayElements(env, input, NULL);

    //compute
    (*env)->ReleaseFloatArrayElements(env, input, _in, 0);
    (*env)->ReleaseFloatArrayElements(env, output, _output, 0);
    return output;
}


voidJava
_com_dsp_matlab_Filters_initialize(JNIEnv *env, jobject thiz)
{

}
```

```
main - Notepad
File  Edit  Format  View  Help
/* your development environment.                                    */
/*                                                                  */
/* This file initializes entry-point function arguments to a default */
/* size and value before calling the entry-point functions. It does */
/* not store or use any values returned from the entry-point functions. */
/* If necessary, it does pre-allocate memory for returned values.   */
/* You can use this file as a starting point for a main function that */
/* you can deploy in your application.                              */
/*                                                                  */
/* After you copy the file, and before you deploy it, you must make the */
/* following changes:                                               */
/* * For variable-size function arguments, change the example sizes to */
/* the sizes that your application requires.                        */
/* * Change the example values of function arguments to the values that */
/* your application requires.                                       */
/* * If the entry-point functions return values, store these values or */
/* otherwise use them as required by your application.              */
/*                                                                  */
/***********************************************************************/
/* Include Files */
#include "rt_nonfinite.h"
#include "LR4_2.h"
#include "main.h"
#include "LR4_2_terminate.h"
#include "LR4_2_emxAPI.h"
#include "LR4_2_initialize.h"

/* Function Declarations */
static void argInit_1xd1024_real32_T(float result_data[], int result_size[2]);
static float argInit_real32_T(void);
static void main_LR4_2(void);
```

Figure 4.52: main.c file located in the folder codegen\lib\LR4_2\examples.

```
void
Java_com_dsp_matlab_Filters_finish(JNIEnv *env, jobject thiz)
{

}
```

6. Navigate to the folder *codegen\lib\LR4_2\ examples* and open the file main.c using Notepad.

7. Copy the part on the top (below Include Files, see Figure 4.52, but not main.h) into the *MATLABNative.c* file of the shell.

8. Next, modify the following statement from

```
Java_com_dsp_matlab_Filters_compute(JNIEnv *env, jobject thiz,
jfloatArray input)
```

to

```
Java_com_dsp_matlab_Filters_compute(JNIEnv *env, jobject thiz,
jfloatArray   input, jint Fs, jint frameSize)
```

To include the sampling frequency and the frame size as input, change the following statement from

```
jfloatArray output = (*env)->NewFloatArray(env, 256);
float *_output = (*env)->GetFloatArrayElements(env, output, NULL);
```

to

```
jfloatArray out = (*env)->NewFloatArray(env, frameSize);
float *_in = (*env)->GetFloatArrayElements(env, input, NULL);
```

These statements define the input and output variables and allocate memory to them.

From the file main.c, copy the following lines that define and initialize the variables and place them in the *MATLABNative.c* file:

```
emxArray_real32_T *con;
int x_size[2]={1,frameSize};
emxInitArray_real32_T(&con, 2);
```

Now copy the following line from the file main.c and place it in the *MATLABNative.c* file:

```
/* Call the entry-point 'LR4_2'. */
LR4_2(_in, x_size, Fs, con);
```

This function calls the function LR4_2, where the three inputs are sampling frequency FS , input data _in , and frame size x_size . The output is returned in con . Add the following line to access the con data and assign it to the out variable:

```
(*env)->SetFloatArrayRegion(env,out,0,frameSize,con->data);
```

Copy this line from main.c and place it in the *MATLABNative.c* file to delete the memory allocated for `con`

```
emxDestroyArray_real32_T(con);
```

Next, release the allocated memory for the input and return the output as follows:

```
(*env)->ReleaseFloatArrayElements(env, input, _in, 0);
return out;
```

Modify the following function from

```
void
Java_com_dsp_matlab_Filters_initialize(JNIEnv *env, jobject thiz)
{
}
```

to

```
void
Java_com_dsp_matlab_Filters_initialize(JNIEnv *env, jobject thiz)
{
    LR4_2_initialize();
}
```

and the following function from

```
void
Java_com_dsp_matlab_Filters_finish(JNIEnv *env, jobject thiz)
{
}
```

to

```
void
Java_com_dsp_matlab_Filters_finish(JNIEnv *env, jobject thiz)
{
    LR4_2_terminate();
}
```

9. To add the frame size as an input setting, navigate to the file *app/src/main/res/xml/prefs.xml* in the project directory under Android Studio and add the following lines:

```
<ListPreference
    android:key="framesize1"
    android:title="Frame Size"
    android:summary="Default: 256"
    android:defaultValue="256"
    android:entries="@array/framesizeOptions"
    android:entryValues="@array/framesizeValues"/>
```

10. Next, navigate to *app/src/main/res/values/arrays.xml* and add the following lines to define frame size options and values for users:

```
<string-array name="framesizeOptions">
    <item>128</item>
    <item>256</item>
    <item>512</item>
    <item>1024</item>
    <item>2048</item>
</string-array>

<string-array name="framesizeValues">
    <item>128</item>
    <item>256</item>
    <item>512</item>
    <item>1024</item>
    <item>2048</item>
</string-array>
```

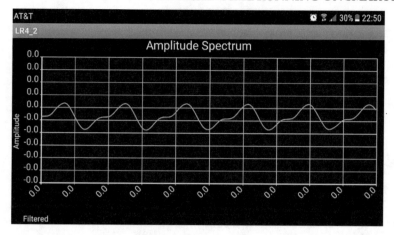

Figure 4.53: LR4_2 filtered microphone output.

11. In the Project Navigator, open the file CMakeLists.txt. Under the comment "#Provides a relative path to your source files(s)", enter the filenames of all the .c and .h files from MATLAB. This will let Android Studio and CMAKE know of these files.

12. Finally, clean project and then run it while having the smartphone connected to your computer. You should be able to see the app screen shown in Figure 4.53 appearing on the smartphone.

13. Note that in order to activate the *Read File* button in the Android app, it is required to enable the permission of storage and microphone for the app (see Figure 4.54).

4.6.4   MODIFYING REAL-TIME SHELL FOR iOS

This subsection covers the modification of the Xcode project for real-time operation. As stated in Section 3.3.2, the MATLAB .c and .h files need to be added to the project and the appropriate C code needs to get inserted. Moreover, if using MATLAB R2016b or later, it is required to add a specific header file into the project folder separately. This header file (tmwtypes.h) can be found in the MATLAB root with the following path:

```
MATLAB root\R2019b\extern\include
```

For iOS, since the layout of the code is different, a different section of the code needs to get modified.

To begin, open the LR4_2 Xcode project and open the file "AudioController.m". This file contains the microphone and C code interaction. For the real-time labs, .caf files or Mac format-

Figure 4.54: Permissions.

ted sound files, are provided. In the Project Navigator, an "Audio Files" group can be seen that reference these files. A MATLAB script named create_signals.m is also provided to regenerate these files in either .wav or .caf format. If it is desired to include other files for processing, they can be added to this group.

Add the following import lines near the top of the file to access the MATLAB generated code:

```
#import <ViewController.h>
#import <LR4_2_initialize.h>
#import <LR4_2_terminate.h>
```

Near the top of the file is the `dealloc()` method, this is called when the AudioController is shutdown. Modify it so that it appears as follows:

```
-(void) dealloc
{
    [[NSNotificationCenter defaultCenter] removeObserver:self];
    free(_filt_data);
    LR4_2_terminate():
}
```

This will safely destroy all the allocated memory in the `init()` method. Add the following code snippet to the `init()` method to allocate data and to initialize the MATLAB routines:

```
_filt_data = malloc(_data_size*sizeof(float));
memset(_filt_data, 0, _data_size*sizeof(float));

LR4_2_initialize();
```

With the memory allocated and the MATLAB files initialized, update the actual data filtering by inserting this function into the file:

```
-(void) filterData:(float*)data withDataLength:(UInt32)length
{
  int bytes_to_process=length;
  int i=0;
  int frameSize=0;

  while(bytes_to_process > 0 )
  {
    frameSize = _frameSize;
    if( bytes_to_process < _frameSize ) {
      frameSize = bytes_to_process;
    }

    emxArray_real32_T *x;
    emxArray_real32_T *con;
    x = emxCreateWrapper_real32_T(data+(i*_frameSize),1,frameSize);
    con = emxCreateWrapper_real32_T(_filt_data+(i*_frameSize),
    1,frameSize+64);

    LR4_2(x, _sampleRate, con);

    BOOL sameptr = _filt_data+(i*_frameSize) == con->data;
    if( !sampeptr ) {
      memcpy(_filt_data+(i*_frameSize), con->data,
        frameSize*sizeof(float));
      free(con->data);
    }
```

```
    bytes_to_process -= _frameSize;
    i++;
  }
}
```

The above block of code uses a different approach when setting up the MATLAB array object. It creates a wrapper object for previously allocated data. Since this function is called repeatedly, time is saved by not allocating and freeing data every block of data.

The EZAudio POD returns data in lengths that are not always equally divisible by the chosen frame size. For this reason, enough of data must be processed until there is less than one frame's worth of data. In this case, whatever remains gets processed.

It should be noted that since the output filtered data are to be stored in the variable " con ," enough memory needs to get allocated. If there is not enough storage, it will allocate more memory and reassign it to a variable. This causes a memory leak. The code after the `LR4_2()` function call handles such situations.

When compiled and run, one should see something similar to the plot shown in Figure 4.55.

For the real-time iOS labs, one can retrieve the output signal file using the share button. This requires turning on Bluetooth to enable AirDrop. To add your own audio file as input, in the Project Navigator, select the "AudioController.h" file to edit it. You will see some `#define` statements toward the top; add your own filename here, e.g.,

```
#define MYFILE [[NSBundle mainBundle] pathForResource:@''filename''
ofType:@"ext"]
```

Then in the "AudioController.m," in the `init()` function, add the following to the _audioList array:

```
_audioList = @[@"NOT_USED",
               SINE,
               SQUARE1,
               SQUARE2,
               SQUARE3,
               SAWTOOTH1,
               SAWTOOTH2,
               SAWTOOTH3,
               CHIRP1,
               CHIRP2,
```

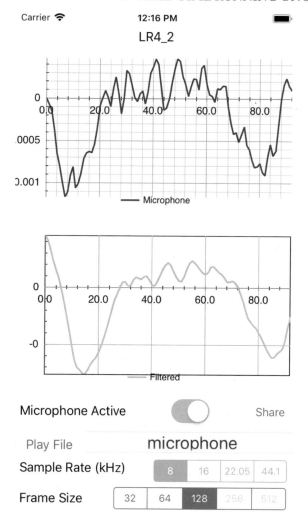

Figure 4.55: LR4_2 microphone input being filtered.

```
                    MYFILE];
```

This makes the code aware of your file. Also, your file in the "Audio Files" group in the Navigator needs to get added. Once the code is recompiled and run, you should be able to see your file in the scroll picker. This method can be done to add your own input file for all of the real-time labs.

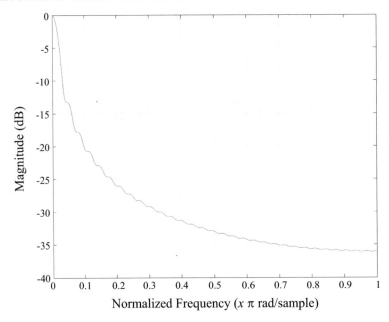

Figure 4.56: Magnitude of system for LR4_3.

4.7 REAL-TIME LABS

LR4_3 – In lab LR4_2, the convolution of a predefined impulse response with voice input is done. Lab LR4_3 is similar except it uses a different filter. Repeat the steps for LR4_2 for Android and iOS in order to run the program. Opening the file LR4_3.m will show the filter is constructed as follows:

```
x2 = 0:2/64:2-2/64;
x2 = x2 / sum(x2);
```

The filter is normalized here so that it will have 0dB gain at its max gain. The magnitude of x2 is shown in Figure 4.56. From this figure, it can be seen that it is a lowpass filter with a small passband bandwidth. When running on an iPhone with a square wave at 100 Hz and 50% duty cycle, the output will appear as shown in Figure 4.57. The input voltage is 0.5 and the output can be seen to be around 0.004.

LR4_4 – The real-time lab LR4_4 allows one to see the distributive and associative property of convolution. Note that it is required to add to the GUI a selection widget of the output plots. As usual, one needs to import the appropriate MATLAB codes.

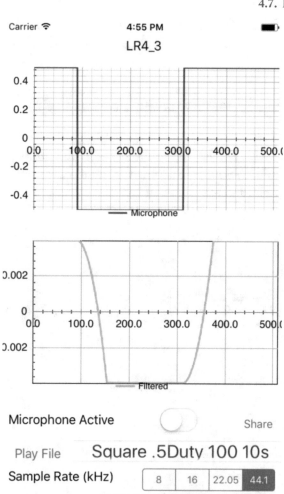

Figure 4.57: iPhone display of LR4_3.

ANDROID STEPS

Let us start from the previous lab LR4_3. Copy the project folder and then update all the referencing of LR4_3 to LR4_4. Make sure to update the app/build.gradle file so that the applicationID has a unique name. This determines whether a new app will appear or a previous one is overwritten.

1. Update the CMakeFiles.txt file with the list of .cpp and .h files generated from MATLAB.

2. If using MATLAB R2016b or later, it is required to add a specific header file into the *jni* folder separately. This header file (`tmwtypes.h`) can be found in the MATLAB root with the following path:

```
MATLAB root\R2019b\extern\include
```

3. To add the selection choice to Android, edit the *arrays.xml* file in the path app/src/main/res/values/. Similar to adding the frame size selection in Section 4.6.3, consider `choiceOptions` and `choiceValues`

```
<resources>
...
<string-array name="choiceOptions">
    <item>(h1*x)*h2</item>
    <item>(h2*x)*h1</item>
    <item>(x*h1)*h2</item>
    <item>(h1*h2)*x</item>
    <item>x*(h1+h2)</item>
    <item>(x*h1)+(x*h2)</item>
</string-array>
<string-array name="choiceValues">
    <item>1</item>
    <item>2</item>
    <item>3</item>
    <item>4</item>
    <item>5</item>
    <item>6</item>
</string-array>
</resources>
```

4. In MatlabNative.c, add the choice parameter to the function signature.

```
Java_com_dsp_matlab_Filters_compute(JNIEnv *env, jobject thiz,
jfloatArray input, jint Fs, jint choice, jint frameSize)
```

This allows one to pass values from Java to the MATLAB C code.

5. Add the new variable as a parameter to the LR4_4 function call.

```
LR4_4(_in, x_size, Fs, choice, y);
```

The variable y will now hold the output of the choice selected. Update the initialize and terminate functions to reflect the new lab number. In the file Filter.java, pass the new choice to the compute function as follows:

```
currentFrame.setFiltered(Filters.compute(currentFrame.getFloats(),
    Settings.Fs, Settings.choice, Settings.blockSize));
```

6. And, lastly, the file Filters.java updates the interface to *MatlabNative.c*. Update it by adding the new parameter,

```
public static native float[] compute(float[] in, int Fs,
int choice, int framesize);
```

These are the typical files that will change from lab to lab. Compile the code and select different outputs.

iOS STEPS

The iOS steps in order to add the plot selection are covered here. Instead of copying the LR4_3 Xcode project, it is better to create a new project. The project initialization was covered previously. After the initialization is done, copy the Main.storyboard, ViewController, and Audio-Controller files from the previous project to the current. This can be done in either Xcode or in the Finder app. This allows modifying the LR4_3 code to the LR4_4 requirements. Again, update all the references of LR4_3 to LR4_4. Add the "Native Code" group and add the .c and .h files from MATLAB. Remember that if using MATLAB R2016b or later, it is required to add a specific header file into the *Native Code* folder separately. This header file (tmwtypes.h) can be found in the MATLAB root with the following path:

```
MATLAB root\R2019b\extern\include
```

Also, add the "Audio Files" group with the included set of .caf files. One can continue to add new picker view option.

1. Add the pickerView from the storyboard view. Some item may have to be re-arranged to fit the new item.

2. Add the code that references the new picker view and also a method that returns the index of the current selection. In ViewController.h, add within the @interface definition.

```
@property (weak, nonatomic) IBOutlet UIPickerView *plotPickerView;
-(NSInteger) getPlotPickerIndex;
```

3. Add the code to ViewController.m. Within the @interface definition, add the following:

```
@property NSArray* plotPickerData;
```

Then, within the `viewDidLoad()` method, add this plot text to be displayed.

```
_plotPickerView.delegate = self;
_plotPickerView.dataSource = self;
_plotPickerData = @[@"(h1*x)*h2",
                    @"(h2*x)*h1",
                    @"(x*h1)*h2",
                    @"(h1*h2)*x",
                    @"x*(h1+h2)",
                    @"(x*h1)+(x*h2)"];
```

The above block of code enables _plotPickerView to call on the ViewController class in order to know how to populate itself and behave when acted upon. This requires defining a few methods which _plotPickerView will call. The numberOfComponentsInPickerView method has already been created and will work as is. This method tells pickerView how many columns of data to expect. In our case, just 1 is used for both pickers. Modify numberOfRowsInMethod to return the correct value based on the component as specified below.

```
-(long)pickerView:(UIPickerView*)pickerView
numberOfRowsInComponet:(NSInteger)component
{
    if( pickerView == _filePickerVIew )
```

```
    {
        return _pickerData.count;
    } else {
        return _plotPickerData.count;
    }
}
```

Consequently, the number of elements in the array created above is returned. Next, when the picker view populates, it needs to know the text for each row. This is done by updating the `titleForRow` method. When called upon, the item from the array is returned for a given row:

```
-(NSString*)pickerView:(UIPickerView*)pckerView titleForRow:
(NSInteger)row forComponent:(NSInteger)component {
    if( pickerView == _filePickerView )  {
        return _pickerData[row];
    } else {
        return _plotPickerData[row];
    }
```

Also in ViewConroller.m, create the method that returns the current index selected. AudioController will use this to know what plot to generate:

```
-(NSInteger) getPlotPickerIndex
{
    Return [_plotPickerView selectedRowInComponent:0];
}
```

4. Last, reference the GUI picker view with the `_plotPickerView` object in the code. This is done in the storyboard view.

   Display the Connections inspector, right most view in the right properties bar. Click and drag from the empty circle to the right of "New Referencing Outlet" to the view controller. A small popup will display three options. Choose the plotPickerView. Figure 4.58 illustrates this. Next, build and run the app.

5. AudioController will need to know which plot one wishes to generate. This value needs to be obtained from the ViewController and passed to the MATLAB function LR4_4(). In AudioController.h, add the method in the @protocol audioUIDelegate as noted below.

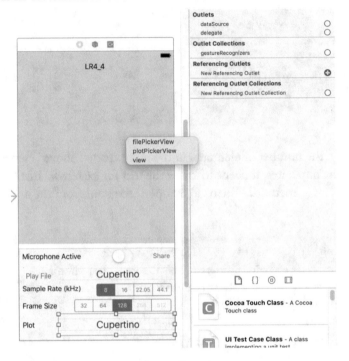

Figure 4.58: Reference the pickerView with an object.

```
@protocal audioUIDelegate
...
-(NSinteger) getPlotPickerIndex;
...
@end
```

Also add a variable to store the value. Within the @interface AudioController section, add the following:

```
@property float plotIndex;
```

Then in AudioController.m, get this value when an action is taken and pass it to the MATLAB function.

Within the function `toggleMicrophone`, add the following near the top:

```
_plotIndex = [_delegate getPlotPickerIndex];
```

In the filterData method, update the function to use the new values. Since the MATLAB code expects the plot index to start at 1, 1 needs to be added to the index. Compile and run the app.

```
LR4_4(x,_sampleRate,_plotIndex, (_plotIndex+1), con);
```

LR4_5 – Real-time lab LR4_5 simulates a real-time RC circuit by using known signals as input or by using live signals from the microphone. The MATLAB code implements the convolution of the input with either Equation (4.13) or (4.15) depending on which selection is made.

ANDROID STEPS

Use LR4_4 as a guide to implement this lab. Remember to update the CMakeLists.txt file to use the appropriate MATLAB code generated for this lab and update codes using the LR4_5 methods.

To change the items to plot, replace the entries defined in the file *app/src/main/res/values/arrays.xml*. It needs to appear like the following when finished:

```
<resources>
...
<string-array name="systemOptions">
    <item>RL Circuit</item>
    <item>RC Circuit</item>

</string-array>
<string-array name="systemValues">
    <item>1</item>
    <item>2</item>
</string-array>
</resources>
```

iOS STEPS

Starting from the LR4_4 lab, a few updates will need to be made. To allow inputs from the user for the resistance, capacitance, and induction values, new GUI elements, variables to store their values, and a way to retrieve the elements from the GUI are needed. Accessing GUI values are briefly mentioned next.

To start, add the GUI elements. One will be able to select which system to plot by modifying the values in the array created in the last lab.

1. Replace the `_plotPickerData` elements with @"RC Circuit" and @"RL Circuit". The rest of the picker view code will work as is. This is modified in the ViewController.m file and the viewDidLoad method.

2. To allow the AudioController to access GUI values, add the following lines to ViewController.h within the @interface section.

```
@property (nonatomic, weak) IBOutlet UITextField *RInput;
@property (nonatomic, weak) IBOutlet UITextField *LInput;
@property (nonatomic, weak) IBOutlet UITextField *CInput;
- (float) getRValue;
- (float) getLValue;
- (float) getCValue;
```

The UITextField pointers allow one to access the text field properties. The get{R,L,C} Value methods allow the AudioController to retrieve the data as floats.

3. Populate the get{R,L,C} Value methods in ViewController.m. These methods pull text values entered, parse the string to a float, then return that value. There is no other error checking.

```
- (float) getRValue
{
   NSString *strVal = _RInput.text;
   return [[NSDecimalNumber decimalNumberWithString:(strVal)]
      floatValue];
}
- (float) getLValue;
{
   NSString *strVal = _LInput.text;
   return [[NSDecimalNumber decimalNumberWithString:(strVal)]
      floatValue];
}
- (float) getCValue;
```

```
{
    NSString *strVal = _CInput.text;
    return  [[NSDecimalNumber decimalNumberWithString:(strVal)]
        floatValue];
}
```

4. In the Main.Storyboard, move up the elements to fit Labels and text inputs for the R, L, C values. Some convenient values to set are the keyboard type= Decimal Pad, correction=no, spell checking=no. As noted before, add the references to appropriate objects as shown in Figure 4.58.

5. Define the member variables and interface methods in AudioController.h. Within the audioUIDelegate protocol section, add these get methods:

```
- (float) getRValue;
- (float) getLValue;
- (float) getCValue;
```

The ViewController is acting as delegate for the AudioController and these methods define the protocol that is needed to be created. Within the @interface section, add these lines to create variables for the R, L, C values:

```
@property (readwrite) float R;
@property (readwrite) float L;
@property (readwrite) float C;
```

6. Next, edit AudioController.m to complete the implementation. As before, edit all the LR4_4 functions to use the LR4_5 functions. In viewDidLoad, initialize the R, L, C variables.

```
_R = 15.0;
_L = 10.0;
_C = 10.0;
```

In the togglePlaybackOutput, add these lines to retrieve the current R, L, C values from the GUI:

```
_plotIndex = [_delegate getPlotPickerIndex];

_R = [_delegate getRValue];
_L = [_delegate getLValue];
_C = [_delegate getCValue];
```

As mentioned before, the AudioController is calling `getRValue`, `getLValue`, and `getCValue` from the ViewController to assign the float values. In the filterData method, update the function to use the new values. Compile and run the app.

```
LR4_5(x,_sampleRate,_plotIndex, _R, _L, _C, con);
```

4.8   REFERENCES

[1] S. Karris. *Signals and Systems with MATLAB Applications*, 2nd ed., Orchard Publications, 2003. 73

[2] J. Buck, M. Daniel, and A. Singer. *Computer Explorations in Signal and Systems Using MATLAB*, 2nd ed., Prentice Hall, 1996. 102

[3] B. Lathi. *Linear Systems and Signals*, 2nd ed., Oxford University Press, 2004.

[4] D. Fannin, R. Ziemer, and W. Tranter. *Signals and Systems: Continuous and Discrete*, 4th ed., Prentice Hall, 1998.

[5] B. Heck and E. Kamen. *Fundamentals of Signals and Systems Using the Web and MATLAB*, 3rd ed., Prentice Hall, 2006.

[6] C. Phillips, E. Riskin, and J. Parr. *Signals, Systems and Transformations*, 3rd ed., Prentice Hall, 2002.

[7] S. Soliman and M. Srinath. *Continuous and Discrete Signals and Systems*, 2nd ed., Prentice Hall, 1998.

[8] M. Roberts. *Signals and Systems*, McGraw-Hill, 2004. 73, 91

CHAPTER 5

Fourier Series

A periodic signal $x(t)$ can be expressed by an exponential Fourier series as follows:

$$x(t) = \sum_{n=-\infty}^{\infty} c_n e^{j\frac{2\pi nt}{T}}, \tag{5.1}$$

where T indicates the period of the signal and c_n's are called Fourier series coefficients, which in general are complex. These coefficients are obtained by performing the following integration:

$$c_n = \frac{1}{T} \int_T x(t) e^{-j\frac{2\pi nt}{T}} dt, \tag{5.2}$$

which possesses the following symmetry properties:

$$|c_{-n}| = |c_n| \tag{5.3}$$

$$\angle c_{-n} = -\angle c_n, \tag{5.4}$$

where the symbol $|.|$ denotes magnitude and $\angle$ phase. Magnitudes of the coefficients possess even symmetry and their phases odd symmetry.

A periodic signal $x(t)$ can also be represented by a trigonometric Fourier series as follows:

$$x(t) = a_0 + \sum_{n=1}^{\infty} a_n \cos\left(\frac{2\pi nt}{T}\right) + b_n \sin\left(\frac{2\pi nt}{T}\right), \tag{5.5}$$

where

$$a_0 = \frac{1}{T} \int_T x(t) dt, \tag{5.6}$$

$$a_n = \frac{2}{T} \int_T x(t) \cos\left(\frac{2\pi nt}{T}\right) dt, \tag{5.7}$$

$$b_n = \frac{2}{T} \int_T x(t) \sin\left(\frac{2\pi nt}{T}\right) dt. \tag{5.8}$$

The relationships between the trigonometric series and the exponential series coefficients are given by

$$a_0 = c_0, \tag{5.9}$$

$$a_n = 2Re\{c_n\}, \tag{5.10}$$

$$b_n = -2Im\{c_n\}, \tag{5.11}$$

$$c_n = \frac{1}{2}(a_n - jb_n), \tag{5.12}$$

where Re and Im denote the real and imaginary parts, respectively.

According to the Parseval's theorem, the average power in the signal $x(t)$ is related to its Fourier series coefficients c_n's, as indicated below:

$$\frac{1}{T}\int_T |x(t)|^2 \, dt = \sum_{n=-\infty}^{\infty} |c_n|^2 . \tag{5.13}$$

More theoretical details of Fourier series are available in signals and systems textbooks, e.g., [1–3].

5.1   FOURIER SERIES NUMERICAL COMPUTATION

The implementation of the integration in Equations (5.6)–(5.8) is achieved by performing summations. In other words, the integrals in (5.6)–(5.8) are approximated by summations of rectangular strips, each of width Δt, as follows:

$$a_0 = \frac{1}{M}\sum_{m=1}^{M} x(m\Delta t), \tag{5.14}$$

$$a_n = \frac{2}{M}\sum_{m=1}^{M} x(m\Delta t)\cos\left(\frac{2\pi mn}{M}\right), \tag{5.15}$$

$$b_n = \frac{2}{M}\sum_{m=1}^{M} x(m\Delta t)\sin\left(\frac{2\pi mn}{M}\right), \tag{5.16}$$

where $x(m\Delta t)$ are M equally spaced data points representing $x(t)$ over a single period T, and Δt indicates the interval between data points such that $\Delta t = \frac{T}{M}$.

Similarly, by approximating the integral in Equation (5.2) with a summation of rectangular strips, each of width Δt, one can write

$$c_n = \frac{1}{M}\sum_{m=-M}^{M} x(m\Delta t)\exp\left(-\frac{j2\pi mn}{M}\right). \tag{5.17}$$

Note that throughout the book, the notations dt, delta, and Δ are used interchangeably to denote the time interval between samples.

Table 5.1: MATLAB functions for generating various waveforms or signals

Waveform Type	MATLAB Function
Square wave	*square*(*T*), *T* denotes period
Triangular wave	*sawtooth* (*T*, *Width*), *Width* = 0.5
Sawtooth wave	*sawtooth* (*T*, *Width*), *Width* = 0
Half-wave rectified sine wave	$\begin{cases} \text{sine}(2 * pi * f * t) & \text{for} \quad 0 \leq t < T/2 \\ 0 & \text{for} \quad T/2 \leq t < T \end{cases}$ *f* = 1/*T* denotes frequency Half period is sine wave and the other half is made zero

5.2   FOURIER SERIES AND ITS APPLICATIONS

In this section, the representation of periodic signals based on Fourier series is considered. Periodic signals can be represented by a linear combination of an infinite sum of sine waves, as expressed by the trigonometric Fourier series representation, Equation (5.5). Periodic signals can also be represented by an infinite sum of harmonically related complex exponentials, as expressed by the exponential Fourier series representation, Equation (5.1). In this lab, both of these series representations are implemented. In particular, the focus is placed on how to compute Fourier series coefficients numerically.

L5.1 FOURIER SERIES SIGNAL DECOMPOSITION AND RECONSTRUCTION

This example helps one to gain an understanding of Fourier series decomposition and reconstruction for periodic signals. The first step involves estimating $x\,(m\Delta t)$ which is a numerical approximation of the periodic input signal. Although programming environments deploy discrete values internally, a close analog approximation of a continuous-time signal can be obtained by using a very small Δt. That is to say, for all practical purposes, when Δt is taken to be very small, an analog signal is simulated. In this example, four input signals are created by using the MATLAB functions listed in Table 5.1.

Open MATLAB, start a new MATLAB script in the HOME panel, and select a *New Script*. Write the MATLAB code to generate these signals: `sin`, `square`, `sawtooth`, and `triangular`. Note the `square` and `sawtooth` MATLAB functions are not supported by the MATLAB Coder toward generating corresponding C codes. For such functions, these functions need to be written from scratch. Use a switch structure to select different types of input waveforms. Set the switch parameter w to serve as the input. Set the amplitude of signal A, period of signal T, and number of Fourier coefficients N as control parameters. Determine the

```matlab
function [XT,x_hat,error,a0,an,bn,max_error,avg_error]=L5_1(dt,A,T,N,w)

    % Generating the input signal
    f=1/T;
    t=0:dt:T-dt;
    X=zeros(1,length(t));
    switch w
    case 0
    X=[A*sin(2*pi*f*(0:dt:T/2-dt)) zeros(1,T/(2*dt))];
    case 1
    % X=A*square(2*pi*f*t);
    X=[A*ones(1,round(T/2/dt)) -A*ones(1,round(T/2/dt))];
    case 2
    % X=A*sawtooth(2*pi*f*t);
    X=2*A/T*(t)-A;
    case 3
    % X=A*sawtooth(2*pi*f*t, 0.5);
    X=[4*A/T*(0:dt:T/2-dt)-A -4*A/T*(T/2:dt:T-dt)+3*A];
    end

    % Creating periodic function
    XT=repmat(X,1,3);
    M=length(X);
    a0=mean(X); % a0 is mean of single period of input
    % Generating Foourier Series Coefficients
    an=zeros(1,N);
    bn=zeros(1,N);
    for i=1:N
      cosfn = cos(2*pi*(i/M)*(1:M));
      an(i) = (2/M)*sum(X.*cosfn);
      sinfn = sin(2*pi*(i/M)*(1:M));
      bn(i) = (2/M)*sum(X.*sinfn);
    end

    % REconstructing the periodic signal
    tind=1:length(XT);
    x_hat=a0*ones(1,length(XT));

    for i=1:N
      x_hat=x_hat+an(i)*cos(2*pi*(i/M)*tind)+bn(i)*sin(2*pi*(i/M)*tind);
    end

    %Calculating the error
    error = abs(XT - x_hat);
    max_error = max(abs(XT - x_hat));
    avg_error = sum(abs(XT - x_hat))/length(XT);
```

Figure 5.1: L5_1 function for the Fourier series signal decomposition and reconstruction example.

Fourier coefficients a_0, a_n, and b_n by using Equations (5.14)–(5.16). Then, reconstruct the signal from its Fourier coefficients using Equation (5.5). Determine the error between the original signal and the reconstructed signal by simply taking the absolute values of $x(t) - \hat{x}(t)$ via the MATLAB function abs. Finally, determine the maximum and average errors by using the

```
Editor - C:\Users\Axa180003\Desktop\MATLAB Examples\Chapter 5\L5_1_testbench.m                    ⊙ x
 L5_1_testbench.m  ✕   +
 1 -    clear;
 2 -    close all
 3 -    clc;
 4
 5 -    Delta=0.001; %Time interval between samples
 6 -    A=1; % Amplitude of input
 7 -    T=2; % Time period of input
 8 -    N=5; % No. of Fourier Series coefficients
 9 -    w=3; %Choice
10
11 -    [XT,x_hat,error,a0,an,bn,max_error,avg_error]=L5_1(Delta,A,T,N,w); %Calling the function which performs operations
12
13      % Verification
14 -    figure(1)
15 -    plot(Delta*(1:length(XT)),XT);grid on;hold on;
16 -    plot(Delta*(1:length(x_hat)),x_hat,'r');
17 -    legend('XT', 'x_hat');
18 -    xlabel('t')
19 -    figure(3)
20 -    plot(an);grid on;hold on;
21 -    plot(bn);
22 -    legend('an','bn')
23 -    figure(2)
24 -    plot(Delta*(1:length(error)),error);grid on;
25 -    title('error')
```

Figure 5.2: L5_1_testbench.

functions `max` and `sum`. Save the script using the name L5_1; see Figure 5.1. Next, write a test script for verification purposes. Open a *New Script*, write your code and save it using the name L5_1_testbench as noted in Figure 5.2.

Run L5_1_testbench for different `A`, `T`, `w`, and `N` values and observe the results. Follow the steps as outlined for L3_1 to generate the corresponding C code and then place it into the shell provided. Figure 5.3 shows the initial screen of the app on an Android smartphone. Enter values for `Delta`, `A`, `T`, and `N`, select the input signal and the desired output to be plotted, then press COMPUTE. Figures 5.5–5.9 show a_n, b_n, periodic signal $x(t)$, reconstructed signal $\hat{x}(t)$, and the error. Note that a_0, Maximum Error, and Average Error are displayed in the main screen of the app.

L5.2 LINEAR CIRCUIT ANALYSIS USING TRIGONOMETRIC FOURIER SERIES

In this example, linear circuit analysis is performed using the trigonometric Fourier series. The ability to decompose any periodic signal into a number of sine waves makes the Fourier series a powerful tool in electrical circuit analysis. The response of a circuit component when a sinusoidal input is applied to its terminals is well known in circuit analysis. Thus, to obtain the response to any periodic signal, one can decompose the signal into sine waves and perform a linear superposition of the sine waves.

Consider a simple RC circuit excited by a periodic input signal, as shown in Figure 5.10.

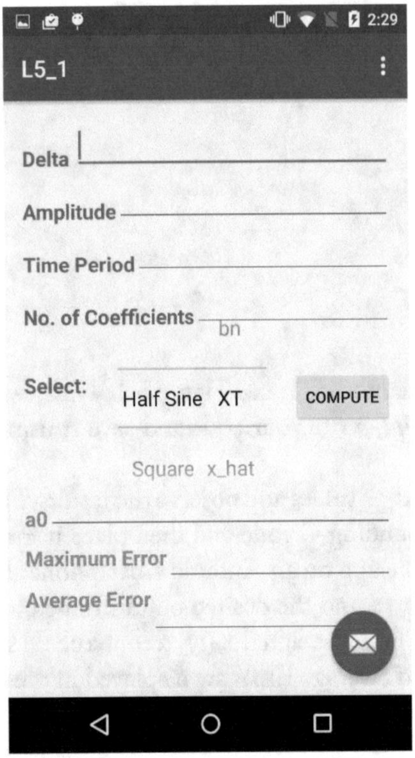

Figure 5.3: Smartphone app screen.

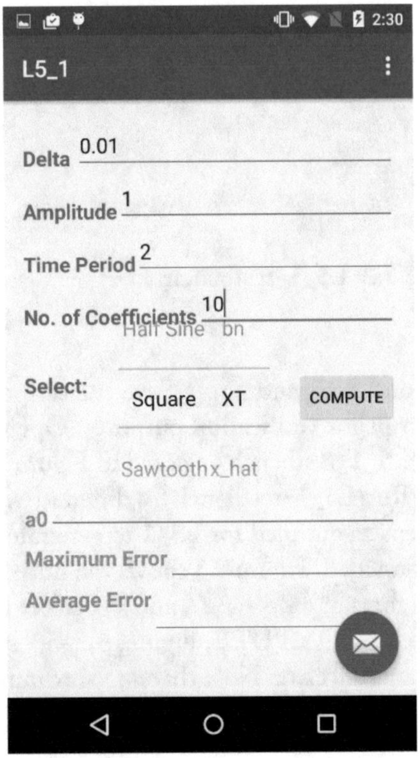

Figure 5.4: Parameter settings.

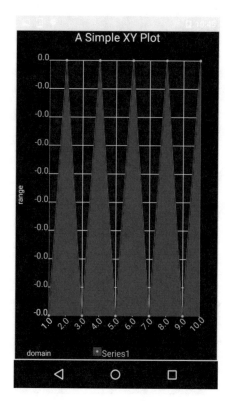

Figure 5.5: Plot of a_n coefficients.

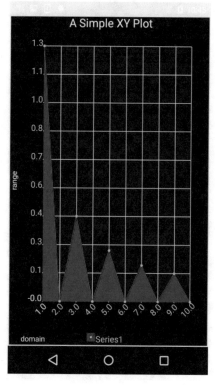

Figure 5.6: Plot of b_n coefficients.

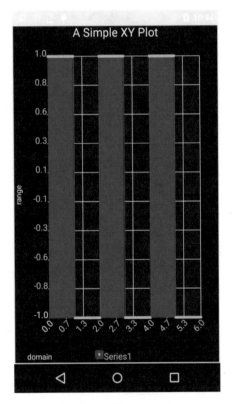

Figure 5.7: Plot of periodic $x(t)$.

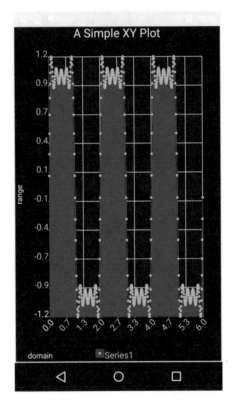

Figure 5.8: Plot of $\hat{x}(t)$ or reconstructed $x(t)$.

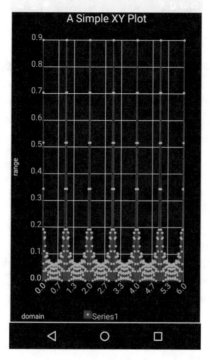

Figure 5.9: Plot of the error between $x(t)$ and $\hat{x}(t)$.

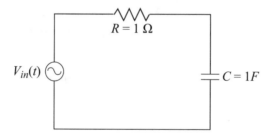

Figure 5.10: RC series circuit with periodic input voltage.

Open MATLAB, start a new MATLAB script in the HOME panel, and select a *New Script*. Write a MATLAB code to determine the Fourier series coefficients of the input voltage signal as discussed in the previous example. Because Fourier series involves the sum of sinusoids, phasor analysis can be used to obtain the output voltage (v_c). Let n represent the number of terms in the Fourier series. By using the voltage divider rule, the output voltage (v_c) can be expressed as [1]:

$$v_{c\_n} = \frac{1/(jn\omega C)}{R + 1/(jn\omega C)} v_{in\_n}. \tag{5.18}$$

Because the sine and cosine components of the input voltage are known, one can easily determine the output by adding the individual output components noting that the circuit is linear. Determine each component of the output voltage by using Equation (5.18). Save the script using the name L5_2, see Figure 5.11. Next, write a test script for verification purposes. Open a New Script, write your verification code and save it using the name L5_2_testbench as shown in Figure 5.12.

Run L5_2_testbench for different resistance, capacitance, A, T, w, and N values and observe the results. Follow the steps as outlined in the lab L3_1 to generate the corresponding C code and then place it into the shell provided. Figure 5.13 shows the app screen on an Android smartphone. Enter values for Delta, A, T, and N, select the input signal and the desired output signal to be plotted, then press COMPUTE. Figures 5.14–5.20 show a_n, b_n, periodic signal $x(t)$, VCcos_m (magnitude of the cosine components), VCcos_a (phase of the cosine components), VCsin_m (magnitude of the sine components), VCsin_a (phase of the sine components), and the error. The parameters a_0 and VCdc are displayed in the main screen of the app.

5.3   LAB EXERCISES

5.3.1   RL CIRCUIT ANALYSIS

Write a MATLAB function to analyze the RL circuit shown in Figure 5.21 using Fourier series.

The input voltage to the circuit is to be either a square wave or a triangular wave with a period $T = 2$ sec. Compute and display the following:

a. the Fourier series coefficients of the input voltage $v(t)$;

b. the current $i(t)$;

c. the root mean square (RMS) value of $v(t)$ using (i) the original waveform and (ii) its Fourier series coefficients (examine the difference); and

d. the average power P_{av} delivered by the voltage source.

RMS VALUE

The RMS value of a periodic function $v(t)$ with period T is given by

$$v_{RMS} = \sqrt{\frac{1}{T} \int_T v^2 dt}. \tag{5.19}$$

Figure 5.11: L5_2 function for circuit analysis with trigonometric Fourier series.

```
Editor - C:\Users\Axa180003\Desktop\MATLAB Examples\Chapter 5\L5_2\L5_2_testbench.m          ⊙ ✕
  L5_2_testbench.m  ✕  +
1 -    clc;clear;
2 -    close all
3 -    Delta=0.001; %Time interval between samples
4 -    A=1; % Amplitude of input
5 -    T=1; % Time period of input
6 -    N=8; % No. of Fourier Series coefficients
7 -    w=2; %Choice
8 -    R=1;C=1; % Resistance and Capacitance values
9      %% Calling the function which performs operations
10 -   [XT,an,bn,a0,VCdc,VCcos_m,VCcos_a,VCsin_m,VCsin_a]=L5_2(Delta,A,T,N,w,R,C);
11     %% Verifying the Results
12 -   figure(1)
13 -   plot(Delta*(1:length(XT)),XT);grid on;
14 -   title('XT')
15 -   xlabel('t')
16 -   figure(2)
17 -   plot(an);grid on;hold on;
18 -   plot(bn);
19 -   legend('an','bn')
20 -   figure(3)
21 -   plot(VCcos_m);grid on;
22 -   title('VCcos-m')
23 -   figure(4)
24 -   plot(VCcos_a);grid on;
25 -   title('VCcos-a')
26 -   figure(5)
27 -   plot(VCsin_m);grid on;
28 -   title('VCsin-m')
29 -   figure(6)
30 -   plot(VCsin_a);grid on;
31 -   title('VCsin-a')
```

Figure 5.12: L5_2_testbench.

Figure 5.13: Smartphone app screen.

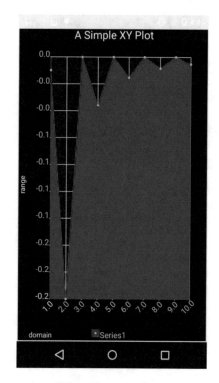

Figure 5.14: Plot of a_n coefficients.

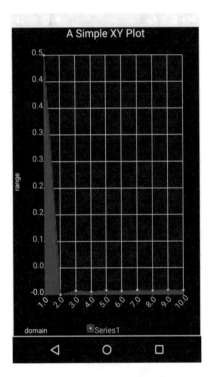

Figure 5.15: Plot of b_n coefficients.

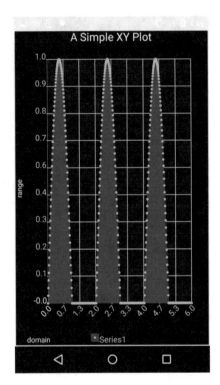

Figure 5.16: Plot of periodic $x(t)$.

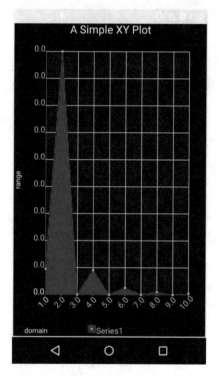

Figure 5.17: Plot of VCcos_m.

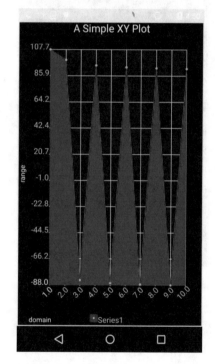

Figure 5.18: Plot of VCcos_a.

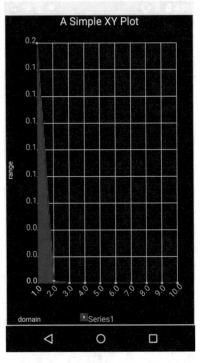

Figure 5.19: Plot of VCSin_m.

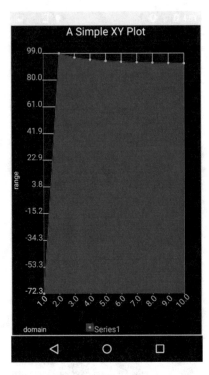

Figure 5.20: Plot of VCSin_a.

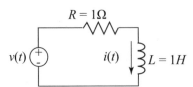

Figure 5.21: RL series circuit with periodic input voltage.

The RMS value of a waveform consisting of sinusoids with different frequencies is equal to the square root of the sum of the squares of the RMS value of each sinusoid. If a waveform is represented by the following Fourier series:

$$v(t) = V_0 + V_1 \sin(\omega_1 t \pm \phi_1) + V_2 \sin(\omega_2 t \pm \phi_2) + \cdots + V_N \sin(\omega_N t \pm \phi_N); \quad (5.20)$$

then, the RMS value V_{RMS} is given by:

$$V_{RMS} = \sqrt{V_0{}^2 + \left(\frac{V_1}{\sqrt{2}}\right)^2 + \left(\frac{V_2}{\sqrt{2}}\right)^2 + \cdots + \left(\frac{V_N}{\sqrt{2}}\right)^2}. \quad (5.21)$$

AVERAGE POWER

The average power of the Fourier series can be expressed as

$$P_{av} = V_0 I_0 + V_{1RMS} I_{1RMS} \cos\phi_1 + V_{2RMS} I_{2RMS} \cos\phi_2 + \cdots. \quad (5.22)$$

5.3.2   DOPPLER EFFECT

Doppler effect means the change in frequency and wavelength of a wave as perceived by an observer moving relative to the wave source. Doppler effect can be demonstrated via time scaling of Fourier series. The observer hears the siren of an approaching emergency vehicle with different amplitudes and frequencies as compared to the original signal. As the vehicle passes by, the observer hears another amplitude and frequency. The reason for the amplitude change (increased loudness) is because of the diminishing distance of the vehicle. The closer it gets, the louder the siren becomes. The reason for frequency (pitch) change is due to the Doppler effect. As the vehicle approaches, each successive compression of the air caused by the siren occurs a little closer than the last one, and the opposite happens when the vehicle moves away. The result is the scaling of the original signal in the time domain, which changes its frequency. When the vehicle approaches, the scaling factor is greater than 1, resulting in a higher frequency, and when it moves away, the scaling factor is less than 1, resulting in a lower frequency. More theoretical aspects of this phenomenon are covered in reference [4].

Define the original siren signal as $x(t)$. When the vehicle approaches, one can describe the signal by

$$x_1(t) = B_1(t)x(at), \quad (5.23)$$

where $B_1(t)$ is an increasing function of time (assuming a linear increment with time) and a is a scaling factor having a value greater than 1. When the vehicle moves away, one can describe the signal by

$$x_2(t) = B_2(t)x(bt), \quad (5.24)$$

where $B_2(t)$ is a decreasing function of time (assuming a linear decrement with time) and b is a scaling factor having a value less than 1.

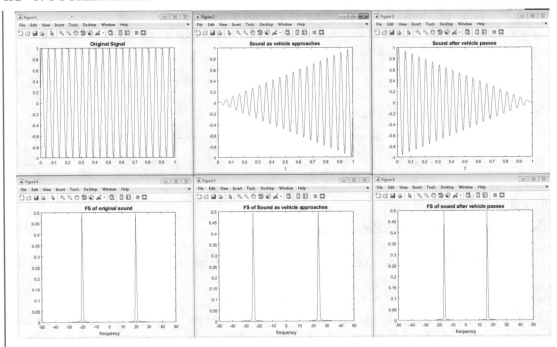

Figure 5.22: Plots of Doppler effect signals.

First, generate a signal and create an upscale and a downscale version of it. Observe the Fourier series for these signals. Set the amplitude and frequency of the original signal and the scaling factors as control parameters. In addition, play the sounds using the function `sound`. Figure 5.22 shows the original sound, and the sound as the vehicle approaches and moves away in both time and frequency domains.

5.3.3   SYNTHESIS OF ELECTRONIC MUSIC

In electronic music instruments, sound generation is implemented via synthesis. Different types of synthesis techniques such as additive synthesis, subtractive synthesis and frequency modulation (FM) synthesis are used to create audio waveforms. The simplest type of synthesis is additive synthesis, where a composite waveform is created by summing sine wave components, which is basically an inverse Fourier series operation. However, in practice, to create a music sound with rich harmonics requires adding a large number of sine waves, which makes the approach computationally inefficient. To avoid adding a large number of sine waves, modulation with addition is used. This exercise involves the design of algorithms used in the Yamaha DX7 music synthesizer, which appeared as the first commercial digital synthesizer.

The primary functional circuit in DX7 consists of a digital sine wave oscillator plus a digital envelope generator. Let us use additive synthesis and frequency modulation to achieve

$y(t)$

Figure 5.23: Additive synthesis.

2 Modulator

1 Carrier

$y(t)$

Figure 5.24: FM synthesis.

$y(t)$

Figure 5.25: Self-modulation.

synthesis with six configurable operators. When one adds together the output of some operators, an additive synthesis occurs, and when one connects the output of one operator to the input of another operator, a modulation occurs.

In terms of block diagrams, the additive synthesis of a waveform with four operators is illustrated in Figure 5.23.

The output for the combination shown in Figure 5.23 can be written as

$$y(t) = A_1 \sin(\omega_1 t) + A_2 \sin(\omega_2 t) + A_3 \sin(\omega_3 t) + A_4 \sin(\omega_4 t). \tag{5.25}$$

Figure 5.24 shows the FM synthesis of a waveform with two operators.

The output for the combination shown in Figure 5.24 can be written as

$$y(t) = A_1 \sin(\omega_1 t + A_2 \sin(\omega_2 t)). \tag{5.26}$$

Other than addition and frequency modulation, one can use feedback or self-modulation in DX7, which involves wrapping back and using the output of an operator to modulate the input of the same operator, as shown in Figure 5.25.

The corresponding equation is

$$y(t) = A_1 \sin(\omega_1 t + y(t)). \tag{5.27}$$

Different arrangements of operators create different algorithms. Figure 5.26 displays the diagram of an algorithm.

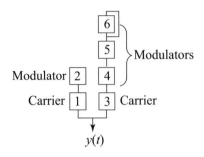

Figure 5.26: Diagram of a synthesis algorithm.

The output for this algorithm can be written as

$$y(t) = A_1 \sin(\omega_1 t + A_2 \sin(\omega_2 t)) + A_3 \sin(\omega_3 t + A_4 \sin(\omega_4 t + A_5 \sin(\omega_5 t + y_6(t)))).$$
(5.28)

With DX7, one can choose from 32 different algorithms. As one moves from algorithm No. 32 to algorithm No. 1, the harmonics complexity increases. In algorithm No. 32, all six operators are combined using additive synthesis with a self-modulator generating the smallest number of harmonics. Figure 5.27 shows the diagram for all 32 combinations of operators. More details on music synthesis and the Yamaha DX7 synthesizer can be found in references [5, 6].

Next, let us explore designing a system with six operators and set their amplitude and frequency as controls. By combining these operators, construct any three algorithms, one from the lower side (for example, algorithm No. 3), one from the middle side (for example, algorithm No. 17) and the final one from the upper side (for example, algorithm No. 30). Observe the output waves in the time and frequency domains by finding the corresponding Fourier series.

5.4 REAL-TIME LABS

LR5_1 – The real-time lab LR5_1 will allow one to examine the Fourier series coefficients of live audio signals. The MATLAB code will decompose the signal into its Fourier series coefficients, then will display the reconstructed signal.

ANDROID STEPS

The steps below indicate the modifications of the lab LR4_5. Copy the LR4_5 directory to LR5_1 and rename LR4_5 to LR5_1. Let us first update the project files before updating the code.

1. Update the applicationId in the app/build.gradle to have LR5_1 in the name instead of LR4_5. Next, copy the MATLAB generated code into the app/src/main/jni directory.

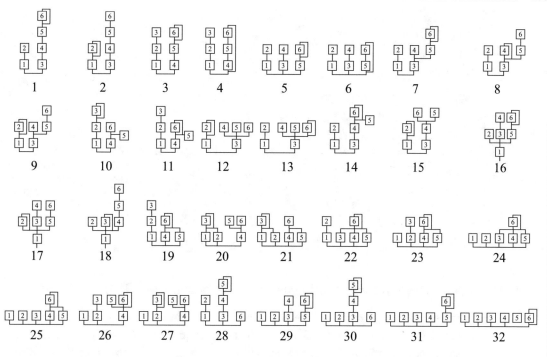

Figure 5.27: Thirty-two algorithms in the Yamaha DX7 music synthesizer.

2. If using MATLAB R2016b or later, it is required to add a specific header file into the *jni* folder separately. This header file (`tmwtypes.h`) can be found in the MATLAB root with the following path:

```
MATLAB root\R2019b\extern\include
```

3. Update the MATLAB generated files in the CMakeLists.txt file. The section in the file should look like the following block of code. The order of the files does not matter.

```
add_library( # Sets the name of the library.
        matlabNative

        # Sets the library as a shared library.
        SHARED

        # Provides a relative path to your source file(s).
```

```
                src/main/jni/LR5_1.c
                src/main/jni/LR5_1.h
                src/main/jni/LR5_1_emxutil.c
                src/main/jni/LR5_1_emxutil.h
                src/main/jni/LR5_1_initialize.c
                src/main/jni/LR5_1_initialize.h
                src/main/jni/LR5_1_terminate.c
                src/main/jni/LR5_1_terminate.h
                src/main/jni/LR5_1_types.h
                src/main/jni/MatlabNative.c
                src/main/jni/rtwtypes.h
                src/main/jni/tmwtypes.h
    )
```

4. Moving onto *AndroidManifest.xml*, update the LR4_5 names to LR5_1.

5. The *app/src/main/res/xml/prefs.xml* file will need to be edited to remove some of the previous options. ListPreference "system1", EditTtextPreferences "resistance1", "inductance1", and "capacitance1" can be removed as they will not be used in this lab. Once these are removed, the following option can be added for the number of coefficients:

```
<EditTextPreference
    android:key="coefficients1"
    android:defaultValue="32"
    android:gravity="left"
    android:inputType="number"
    android:summary="Default: 32"
    android:title="No. of Coefficients" />
```

6. The code update will contain small updates for many files. First, open the fileapp/src/mian/java/com/dsp/matlab/BetteryXYSeries.java. This change will update the graph to display the sample number instead of time. In the method getX(int index), make the return statement return only the index as follows:

```
Public Number getX(int index)
{
        return index;
```

```
}
```

7. A similar change to specify the limits of the graph is needed in app/src/main/java/com/dsp/matlab/DataGraphActivity.java. Update the following line in the onCreate method as follows:

```
DataPlot.setDomainBoundaries(0, Settings.blockSize,
BoundaryMode.Fixed);
```

This will ensure having the right x-axis limits.

8. Update the file app/src/main/java/com/dsp/matlab/Filter.java by replacing the Settings.Fs argument as follows:

```
currentFrame.setFiltered(Filters.compute(currentFrame.getFloats(),
Settings.coefficients, Settings.blockSize));
```

9. Now in the file app/src/main/java/com/dsp/matlab/Filters.java, update the compute method as follows:

```
public static native float[] compute( float[] in, int coefficients,
int frameSize );
```

10. To make sure the coefficient input are applied to the settings, update the updateSettings() method in app/src/main/java/com/dsp/matlab/RealTime.java by adding the following line

```
Settings.setCoefficients(Integer.parseInt(preferences.getString
("coefficients1", "32")));
```

11. Update app/src/main/java/com/dsp/matlab/Settings.java to store and retrieve the coefficients choice by adding the following:

```
public static int coefficients = 32;
public static void setCoefficients( int coefficients1 )
{ coefficients = coefficients1; }
```

12. Finally, update the java to C interface file app/src/main/jni/MatlabNative.c as noted below.

```
#include <jni.h>
#include <stdio.h>
#include "LR5_1.h"
#include "LR5_1_terminate.h"
#include "LR5_1_intialize.h"
#include "tmwtypes.h"

jfloatArray
Java_com_dsp_matlab_Filters_compute(JNIEnv *env, jobject this,
jfloatArray, input, jfloat coefficeints, jint frameSize)
{
    jfloatArray output = (*env)->NewFloatArray(env, frameSize);
    float *_output =
                (*env)->GetFloatArrayElements(env, output, NULL);
    float *_in = (*env)->GetFloatArrayElements(env, input, NULL);
    //compute
    int x_size[2] = {1, frameSize};
    int x_hat_size[2];
    LR5_1(_in, x_size, coefficients, _ouptut, x_hat_size);

    (*env)->ReleaseFloatArrayElements(env, input, _in, 0);
    (*env)->ReleaseFloatArrayElements(env, output,_output, 0);

    return output;
}

Void Java_com_dsp_matlab_Filters_initialize(JNIENV *env,
jobject this) { LR5_1_initialize(); }

Void Java_com_dsp_matlab_Filters_finish(JNIENV *env,
jobject this) { LR5_1_terminate(); }
```

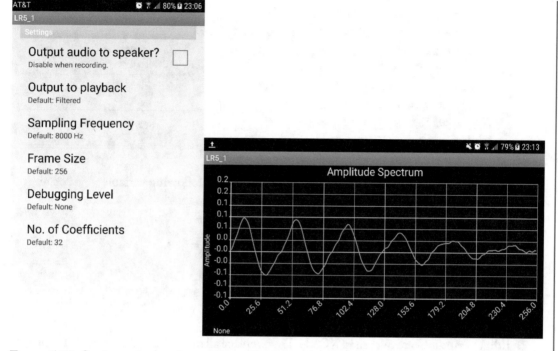

Figure 5.28: Settings view and waveform of input audio.

One can see that all the LR4_5 references are changed to LR5_1. The compute method signature is updated to accept all the arguments. The LR4_5 call is replaced by LR4_5 and the appropriate arguments.

If everything is done correctly, you should see the outputs as shown in Figure 5.28.

iOS STEPS

As done in Chapter 4, for iOS it is easier to start a new project without copying and editing the project references. Therefore, a new project is created as discussed earlier. After the project is created, copy the AudioController.m, AudioController.h, Main.storyboard, ViewController.h, and ViewController.m files from LR4_5 to the appropriate location in the LR5_1 project. Right away, the code can get updated.

1. Edit the AudioController.h file in XCode. Replace the LR4_5 references to LR5_1. From the `AudioDelegate` protocol, remove the items `getRValue`, `getLValue`, `getCValue`, `getPlotPickerIndex`, and the `didStopPlaying`. Add a `getNumCoeffs` method. The protocol section should then look like the following:

```
@protocol AudioDelegate
-(void) updateGraphData:(float*)data withFilterData:
(float*)filtered_data numPoints:(int)numPoints;
-(NSInteger) getPickerIndex;
-(int) getNumCoeffs;
-(void)stoppedPlaying;
-(void didStopRecording;
@end
```

Within the AudioController interface section, add the following variable `numCoeffs`:

```
@interface AudioController ...
...
@property int numCoeffs
...
@end
```

2. Edit the AudioController.m in XCode. First, replace all of the LR4_5 references to LR5_1, then update the include lines, initialize and terminate functions. In the init method, set the default value for `numCoeffs` as follows:

```
...
_frameSize=128;
_numCoeffs =32;
...
```

3. In the `togglePlaybackOutput` method, one needs to get the `numCoeff` value from the AudioController. Add the following call to the delegate:

```
...
NSInteger ind = [_delegate getPickerIndex];
_numCoeffs = [_delegate getNumCOeffs];
...
```

4. Update the LR5_1 method call in filterData. The call should be reduced to three arguments.

```
LR5_1( x -> data, x->size, _numCoeffs, con->data, con->size );
```

5. Moving on to the ViewController.h file, one needs to add the selection of the number of coefficients. Within the `ViewController` interface, remove the RInput, LInput, CInput TextField objects. Also, remove the methods `getRValue`, `getLValue`, `getCValue`, and `getPlotPickerIndex`. Furthermore, remove the `plotPickerView` pointer. A pointer is added to a `UISegmentedControl` and a method is added to handle its changes as follows:

```
@property (nonatomic,weak) IBOutlet UISegmentedControl
                                    *coeffControl;
-(IBAction) updateSelectedNumCoeffs: (id)sender;
```

6. In the AudioController.m file, remove all the previous items and add the plot picker view. References to the `_plotPickerView`, `_RInput`, `_LInput`, `_CInput`, `getRValue`, `getLValue`, `getCvalue`, and `getPlotPickerIndex` methods can be removed. Also, the `numberOfRowsInComponent` and `titleForRow` methods can be simplified to handle the view _filePickerView only.

7. Next, add a few methods. Implement the `updateSelectedNumCoeffs` and `getNumCoeffs methods`. The former is the call back function whenever a new value is selected for the number of coefficients. The latter is called by the `AudioController` to retrieve the latest value.

```
-(IBAction)updateSelectedNumCoeffs:(id)sender
{
    NSInteger index = [sender selectedSegmentIndex];
    NSString *numCoeffStr = [sender titleforSegmentAtIndex:index];
    _numCoeffs = [[NSDecimalNumber decimalNumberWithString:
    (numCoeffStr)] intValue];
}

-(int)getNumCoeffs
{
    return _numCoeffs;
}
```

8. Update the main storyboard to add a selection of the coefficients. In this example, a segmented control is used. In the space where there is a picker view and R, L, C text edit boxes, place the new segmented control. In the Attributes Inspector (right panel), add three segments and assign the titles of "32", "64", and "128". These values will show up as selections similar to how the sample rate selections show up.

9. In the Connections Inspector, associate the valueChanged action with the updateSelectedNumCoeffs method. This is done by dragging the empty circle to the right of the "Value Changed" event over to the storyboard. A popup will list available IBActions. Choose updateSelectedNumCoeffs. This was covered before in Chapter 3.

10. If using MATLAB R2016b or later, it is required to add a specific header file into the *Native Code* folder separately. This header file (`tmwtypes.h`) can be found in the MATLAB root with the following path:

```
MATLAB root\R2019b\extern\include
```

Compile and run your app. Figure 5.29 shows an example screen for this lab.

LR5_2 – The real-time lab LR5_2 again computes the Fourier coefficients of the input signal while being able to choose the output signal. The user will be able to select from the cos coefficients a_n, sin coefficients b_n, mean of the signal, and the magnitude and angle of the sin and cosine components as output of an RC circuit.

ANDROID STEPS

These steps modify the LR5_1 lab. Copy the LR5_1 directory to LR5_2 and rename the rest of the files from LR5_1 to LR5_2. Update the project files before updating the code.

1. Update the applicationId in app/build.gradle to have LR5_2 in the name instead of LR5_1. Next, copy the MATLAB generated code into the directory app/src/main/jni.

2. If using MATLAB R2016b or later, it is required to add a specific header file into the *jni* folder separately. This header file (`tmwtypes.h`) can be found in the MATLAB root with the following path:

```
MATLAB root\R2019b\extern\include
```

3. Update the MATLAB generated files in the file CMakeLists.txt. The section in the file should look like the following block of code. The order of the files does not matter.

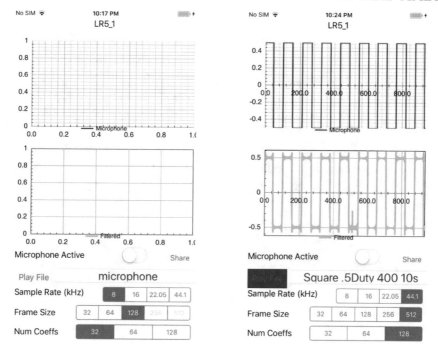

Figure 5.29: Example screen shots for LR5_1.

```
add_library( # Sets the name of the library.
        matlabNative

        # Sets the library as a shared library.
        SHARED

        # Provides a relative path to your source file(s).
        src/main/jni/LR5_2.c
        src/main/jni/LR5_2.h
        src/main/jni/LR5_2_data.c
        src/main/jni/LR5_2_data.h
        src/main/jni/LR5_2_emxAPI.c
        src/main/jni/LR5_2_emxAPI.h
        src/main/jni/LR5_2_emxutil.c
        src/main/jni/LR5_2_emxutil.h
        src/main/jni/LR5_2_initialize.c
```

```
src/main/jni/LR5_2_initialize.h
src/main/jni/LR5_2_terminate.c
src/main/jni/LR5_2_terminate.h
src/main/jni/LR5_2_types.h
src/main/jni/MatlabNative.c
src/main/jni/rt_defines.h
src/main/jni/rt_nonfinite.c
src/main/jni/rt_nonfinite.h
src/main/jni/rtGetInf.c
src/main/jni/rtGetInf.h
src/main/jni/rtGetNaN.c
src/main/jni/rtGetNaN.h
src/main/jni/rtwtypes.h
src/main/jni/tmwtypes.h
)
```

4. Moving onto *AndroidManifest.xml*, update the LR5_1 names to LR5_2.

5. The *app/src/main/res/xml/prefs.xml* file will need to be edited to add the resistance, capacitance, frequency, and output settings. The number of coefficients option can be reused. Add the following options:

```xml
<EditTextPreference
    android:key="resistance1"
    android:defaultValue="1"
    android:gravity="left"
    android:inputType="number"
    android:summary="Default: 1"
    android:title="Resistance (in Ohms)" />
<EditTextPreference
    android:key="capacitance1"
    android:defaultValue="1"
    android:gravity="left"
    android:inputType="number"
    android:summary="Default: 1"
    android:title="Capacitance (in Farad)" />
<EditTextPreference
    android:key="frequency1"
```

```
        android:defaultValue="100"
        android:gravity="left"
        android:inputType="number"
        android:summary="Default: 100"
        android:title="Frequency of input signal" />
<EditTextPreference
        android:key="system1"
        android:title="What to output?"
        android:summary="Default: an"
        android:defaultValue="0"
        android:gravity="left"
        android:entries="@array/systemOptions"
        android:entryValues="@array/systemValues"/>
```

As mentioned earlier, @array/systemOptions and @array/systemValues are defined in the *app/src/main/res/values/arrays.xml* file. Update these next. Simply add within the xml file the following:

```
<string-array name="systemOptions">
    <item>an</item>
    <item>bn</item>
    <item>a0</item>
    <item>VCdc</item>
    <item>VCcos_m</item>
    <item>VCcos_a</item>
    <item>VCsin_m</item>
    <item>VCsin_a</item>
</string-array>
<string-array name="systemValues">
    <item>0</item>
    <item>1</item>
    <item>2</item>
    <item>3</item>
    <item>4</item>
    <item>5</item>
    <item>6</item>
    <item{>7</item>
```

```
</string-array>
```

6. The code update will contain small updates for many files. Update app/src/main/java/com/dsp/matlab/Filter.java to update the function calls. Add the resistance, capacitance, frequency, and display choice as follows:

```
currentFrame.setFiltered(Filters.compute(currentFrame.getFloats(),
Settings.coefficients, Settings.resistance, Settings.capacitance,
Settings.frequency, Settings.system, Settings.blockSize));
```

7. Now in app/src/main/java/com/dsp/matlab/Filters.java, update the compute method as follows:

```
public static native float[] compute( float[] in, int coefficients,
float R, float C, float f, float choice, int frameSize );
```

8. Apply the coefficient input to the settings, update the `updateSettings()` method in app/src/main/java/com/dsp/matlab/RealTime.Java and add the following lines within it:

```
Settings.setResistance(Integer.parseInt(preferences.getString
("resistancd1", "1")));
Settings.setCapacitance(Integer.parseInt(preferences.getString
("capacitance1", "1")));
Settings.setFrequency(Integer.parseInt(preferences.getString
("frequency1", "100")));
Settings.setSystem(Integer.parseInt(preferences.getString
("system1", "0")));
```

9. Update app/src/main/java/com/dsp/matlab/Settings.java to store and retrieve the settings added as follows:

```
public static float resistance = 1;
public static float capacitance = 1;
public static float frequency = 100;
```

```
public static float system = 0;

public static void setResistance(float resistance1)
  { resistance=resistance1; }
public static void setCapacitance(float capacitance1 )
  { capacitance = capacitance1; }
public static void setFrequency(float frequency1)
  { frequency = frequency1; }
public static void setSystem(float system1){ system = system1; }
```

10. Finally, update the app/src/main/jni/MatlabNative.c interface file. This file is the interface from Java to native C. The entire file is listed below.

```
#include <jni.h>
#include <stdio.h>
#include "rt_nonfinite.h"
#include "LR5_2.h"
#include "LR5_2_terminate.h"
#include "LR5_2_emxAPI.h"
#include "LR5_2_intialize.h"

jfloatArray
Java_com_dsp_matlab_Filters_compute(JNIEnv *env, jobject this,
jfloatArray, input, jfloat N, jfloat R, jfloat C, jfloat f,
jfloat choice, jint frameSize)
{
    jfloatArray output = (*env)-${>}$NewFloatArray(env, frameSize);
    float *_in = (*env)->GetFloatArrayElements(env, input, NULL );

    //compute
    emxArray_real32_T *v;
    int X_size[2] = {1, frameSize};
    emxInitArray_real32_T(&V,2);

    LR5_2(_in, X_size, N, R, C, f, choice, V);

    (*env)->SetFloatArrayRegion(env,output,0,frameSize, V->data);
```

```
    emxDestroyArray_real32_T(V);
    (*env)->ReleaseFloatArrayElements(env, input, _in, 0);
    return output;
}


Void Java_com_dsp_matlab_Filters_initialize
    (JNIENV *env, jobject this){ LR5_2_initialize(); }


Void Java_com_dsp_matlab_Filters_finish(JNIENV *env, jobject this)
{ LR5_2_terminate(); }
```

One can see that all the LR5_1 references are updated to LR5_2. The compute method signature is updated to accept all the arguments. The LR5_1 call is replaced by LR5_2 and the appropriate arguments.

If everything is done correctly, you should be able to see the outputs as shown in Figure 5.30.

iOS STEPS

Start a new project as discussed in the previous chapters. Copy the AudioController.m, AudioController.h, Main.storyboard, ViewController.h, and ViewController.m files from LR5_1 to the appropriate location in the LR5_2 project. The iOS project is limited to the inputs that contain single frequency tones. The lab uses the frequency for some of the calculations. To provide a better representation, limits are used for the input files as well as the microphone input.

1. Edit the AudioController.h file in XCode. Replace the LR5_1 references to LR5_2. In the `AudioDelegate` protocol, add the methods `getFrequency`, `getPlotPickerIndex`, `getResistance`, and `getCapacitance`. `getNumCoeffs` can be re-used. The protocol block will look like the following. The `updateGraphData` method changes since the displays are of different lengths.

```
@protocol AudioDelegate
-(void) updateGraphData:(float*)data withDataPoints:
    (int)numDataPoints
    withFilterData:(float*)filtered_data numPoints:(int)
    numFilteredPoints;
-(NSInteger) getPickerIndex;
-(NSInteger) getPlotPickerIndex;
```

Figure 5.30: Settings view (left); output of coefficients (right).

```
-(float) getResistance;
-(float) getFrequency;
-(float) getCapacitance;
-(int) getNumCoeffs;
-(void)didStopPlaying;
-(void didStopRecording;
@end
```

Within the AudioController interface section, add the following:

```
@interface AudioController ...
...
@property NSInteger plotIndex;
```

```
@property NSInteger playbackIndex;
@property float resistance;
@property float capacitance;
@property float frequency;

@property NSInteger filtered_size;
...
@end
```

2. Edit the AudioController.m in XCode. First, replace all of the LR5_1 references to LR5_2, then updating the include lines, initialize and terminate functions. In the init method, set the default values for the new variables as follows:

```
...
_sampleRate=44100.0;
_frameSize = 128;
_numCoeffs =32;
_frequency= 100;
_audioList = @[SQUARE1,
               SQUARE2,
               SQUARE3,
               SAWTOOTH1,
               SAWTOOTH2,
               SAWTOOTH3];
...
```

3. In the `togglePlaybackOutput` method, one needs to get the new values from Audio-Controller by add a call to the delegate as follows:

```
...
_playbackIndex = [_delegate getPickerIndex];
_plotIndex = [_delegate getPlotPickerIndex];
_resistance = [_delegate getResistance];
_capacitance = [_delegate getCapacitance];
_numCoeffs = [_delegate getNumCOeffs];
_frequency = [_delegate getFrequency];
...
```

4. To handle a different output size, update the EZAudioPlayerDelegate method. A couple of the methods will change from

```
[self.filtRecorder appendDataFromBufferList:filteredBufferList
withBufferSize:bufferSize];
...
dispatch_async( dispatch_get_main_queue(), ^{
    [weakSelf.delegate updateGraphData:buffer[0]
    withFilteredData:_filt_data numPoints:bufferSize];
});
```

to

```
[self.filtRecorder appendDataFromBufferList:filteredBufferList
withBufferSize:bufferSize];
...
dispatch_async(dispatch_get_main_queue(),  ^{
    [weakSelf.delegate updateGraphData:buffer[0] withDataPoints:
        bufferSize withFilteredData:_filt_data
        numFilteredPoints:_filtered\_size];
});
```

5. In the previous labs, the input into blocks of the specified coefficient size was processed. In this lab, however, the output is only one block of coefficient size. Therefore, remove the loop to process the data in blocks and process all the data in one frame as follows:

```
-(void) filterData:(float*)data withDataLength:(UInt32)length
{
    emxArray_real32_T *x = emxCreateWrapper_real32_T(data,1,length);
    emxArray_real32_T *con = emxCreateWrapper_real32_T(_filt_data,
                            1, _numCoeffs );
    LR5_2( x, _numCoeffs, _resistance, _capacitance, _frequency,
            _plotIndex, con );

    BOOL sameptr = _filt_data == con->data;
    if( !sameptr )
    {
```

```
        memcpy(_filt_data, con->data, _numCoeffs*sizeof(float));
        free(con->data);
    }
    _filtered_size = _numCoeffs;
}
```

6. Moving on to the ViewController.h file, add the get methods to meet the delegate protocol within the interface definition as follows:

```
-(float) getResistance;
-(float) getFrequency;
-(float) getNumCoeffs;
-(float) getCapacitance;
```

7. In the AudioController.m file, add the UITexfield pointers to be able to retrieve the text input data. Also, add a variable to hold the size of the filtered data since it is different than the input audio data. In addition, add in the picker controls to select the desired plot. Some of the updates are similar to the previous labs.

```
@property (atomic, readwrite) int filteredLen;
@property NSArray* plotPickerData;
@property (nonatomic, weak) IBOutlet UITextField *numCOeffsInput;
@property (nonatomic, weak) IBOutlet UITextField *resistanceInput;
@property (nonatomic, weak) IBOutlet UITextField *capacitanceInput;
@property (nonatomic, weak) IBOutlet UITextField *frequencyInput;
```

8. Now, implement these methods and update the others:

```
-(void)viewDidLoad {
...
// set some default values
_dataLen = 0;
_filteredLen = 0;
_plotPickerView.delegate = self;
_plotPickerView.dataSource = self;
```

```
//Array of values displayed in picker
_plotPickerData = @[@"an",
                    @"bn",
                    @"a0",
                    @"VCdc",
                    @"VCCos_m",
                    @"VCCos_a",
                    @"VCSin_m",
                    @"VCSin_a" ];
    [self addDoneToInputs];
}
...
-(NSUInteger)numberOfRecordsForPlot:(CPTPlot *)plot
{
    If( [plot.identifier isEqual:@"Microphone"])
    {
        return (NSUInteger)_dataLen;
    }
    else {
        return (NSUInteger)_filteredLen;
    }
}
...
-(float)getNumCoeffs
{
    return [[NSDecimalNumber decimalNumberWIthString:
                _numCoeffsInput.text] floatValue];
}
-(float) getResistance
{
    return [[NSDecimalNumber decimalNumberWithString:
                _resistanceInput.text] floatValue];
}
-(float) getCapacitance
{
    return [[NSDecimalNumber decimalNumberWithString:
                _capacitanceInput.text] floatValue];
}
```

```objc
-(float) getFrequency
{
   return [[NSDecimalNumber decimalNumberWithString:
             __frequencyInput.text] floatValue];
}
...
-(void) updateGraphData:(float*)data withDataPoints:
        (int)numDataPoints withFilteredData:float*)filtered_data
        numFilteredPoints:(int)numFilteredPoints
{
   _data = data;
   memcpy(_filtered_data, filtered_data, sizeof(float)
   *numFilteredPoints);
   _dataLen = numDataPoints;
   _filteredLen = numFilteredPoints;
}
...
-(NSInteger) getPlocPickerIndex
{
   return [_plotPickerView selectedRowInComponent:0];
{
//Handle new picker view
-(long)pickerView:(UIPickerView*)pickerView
numberOfRowsInComponent: (NSInteger)component
{
   long rv = 0;
   if( pickerView == _pickerView )
   {
      rv = _pickerData.count;
   }
   else {
      rv = _plotPickerData.count;
   }
   return rv;
}
-(NSString*)pickerView:(UIPickerView*)pickerView titleForRow:
            (NSInteger)rowforComponent:(NSInteger)component
{
```

```
    NSString* rv;
    if( pickerView == _pickerView )
    {
        rv = _pickerData[row];
    {
    else {
        rv = \_plotPickerData[row];
    }
    return rv;
}
```

9. To update the storyboard, first remove the selection of the sampling rate and block size. This will leave space to add resistance, capacitance, and frequency text inputs. Refer to the previous labs as how to setup the text inputs. Lastly, add the picker view. Do not forget to assign the references to variables created earlier. This was shown earlier in Chapter 3.

Compile and run your app. Figure 5.31 shows an example screen of this lab.

5.5   REFERENCES

[1] J. Buck, M. Daniel, and A. Singer. *Computer Explorations in Signal and Systems Using MATLAB*, 2nd ed., Prentice Hall, 1996. DOI: 10.1109/ICASSP.2015.7178293. 136, 144

[2] B. Lathi. *Linear Systems and Signals*, 2nd ed., Oxford University Press, 2004.

[3] D. Fannin, R. Ziemer, and W. Tranter. *Signals and Systems: Continuous and Discrete*, 4th ed., Prentice Hall, 1998. 136

[4] M. Schetzen. *Airborne Doppler Radar: Applications, Theory, and Philosophy*, AIAA Publisher, 2006. 151

[5] D. Benson. *Music: A Mathematical Offering*, Cambridge University Press, 2006. 154

[6] DX7 Digital Programmable Algorithm Synthesizer, YAMAHA Manual, 1999. 154

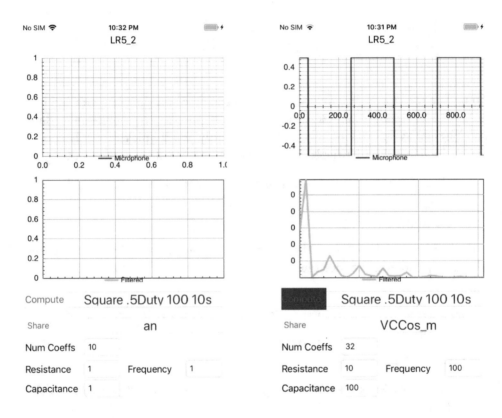

Figure 5.31: Example screen shots for LR5_2.

CHAPTER 6

Continuous-Time Fourier Transform

In this chapter, the continuous-time Fourier transform (CTFT), often referred to as Fourier transform, is computed numerically. This transform is then used to solve linear systems. Also, noise cancellation and amplitude modulation are examined as applications of Fourier transform.

6.1 CTFT AND ITS PROPERTIES

The continuous-time Fourier transform (CTFT) (commonly known as Fourier transform) of a signal $x(t)$ is expressed as

$$X(\omega) = \int_{-\infty}^{\infty} x(t)e^{-j\omega t}\,dt. \tag{6.1}$$

The signal $x(t)$ can be recovered from $X(\omega)$ via this inverse transform equation

$$x(t) = \frac{1}{2\pi} \int_{-\infty}^{\infty} X(\omega)e^{j\omega t}\,d\omega. \tag{6.2}$$

Some of the properties of CTFT are listed in Table 6.1.

Refer to the signals and systems textbooks, e.g., [1–3], for more theoretical details on this transform.

6.2 NUMERICAL APPROXIMATIONS OF CTFT

Assuming that the signal $x(t)$ is zero for $t < 0$ and $t \geq T$, one can approximate the CTFT integration in Equation (6.1) as follows:

$$\int_{-\infty}^{\infty} x(t)e^{-j\omega t}\,dt = \int_{0}^{T} x(t)e^{-j\omega t}\,dt \approx \sum_{n=0}^{N-1} x(n\tau)e^{-j\omega n\tau}\tau, \tag{6.3}$$

where $T = N\tau$ and N is an integer. For sufficiently small τ, the above summation provides a close approximation of the CTFT integral. The summation $\sum_{n=0}^{N-1} x(n\tau)e^{-j\omega n\tau}$ is widely used

Table 6.1: CTFT properties

Properties	Time Domain	Frequency Domain
Time shift	$x(t - t_0)$	$X(\omega)e^{-j\omega t_0}$
Time scaling	$x(at)$	$\dfrac{1}{|a|} X\left(\dfrac{\omega}{a}\right)$
Linearity	$a_1 x_1(t) + a_2 x_2(t)$	$a_1 X_1(\omega) + a_2 X_2(\omega)$
Time convolution	$x(t) * h(t)$	$X(\omega)H(\omega)$
Frequency convolution	$x(t)h(t)$	$X(\omega) * H(\omega)$

in digital signal processing, and MATLAB has a built-in function for it called `fft`. In a .m file, if N samples $x(n\tau)$ are stored in a vector x, then the function call

```
>> xw=tau*fft(x)
```

computes

$$X_\omega(k+1) = \tau \sum_{n=0}^{N-1} x(n\tau)e^{-j\omega_k n\tau} \qquad 0 \le k \le N-1, \tag{6.4}$$

$$\approx X(\omega_k),$$

where

$$\omega_k = \begin{cases} \frac{2\pi k}{N\tau} & 0 \le k \le \frac{N}{2} \\ \frac{2\pi k}{N\tau} - \frac{2\pi}{\tau} & \frac{N}{2}+1 \le k \le N-1, \end{cases} \tag{6.5}$$

with N assumed to be even. The `fft` function returns the positive frequency samples before the negative frequency samples. To place the frequency samples in the right order to compute the inverse transform, the function `fftshift` can be used as indicated below:

```
>>xw=fftshift(tau*fft(x))
```

Note that $X(\omega)$ is a vector (actually, a complex vector) of dimension N. $X(\omega)$ is complex in general despite the fact that $x(t)$ is real valued. The magnitude of $X(\omega)$ is computed using the function `abs` and its phase using the function `angle`.

6.3   EVALUATING PROPERTIES OF CTFT

The example covered in this section provides an implementation of CTFT and its properties. As mentioned earlier, implementation is carried out only in a discrete fashion. Thus, to get a continuous-time representation of a signal, a very small value of time increment Δt is used. For example, $\Delta t = 0.001$ means there are 1000 samples in 1 s, generating a good simulation of a low-frequency continuous signal. However, when working with very high-frequency signals, Δt should be decreased further to ensure there are enough samples to adequately simulate them in a continuous fashion over a specified duration.

L6.1 EVALUATING PROPERTIES OF CTFT EXAMPLE

Let us write a MATLAB code to create two input signals x1 and x2. Set up a case structure by using **mode3** and **mode4** for controlling x1 and x2 type, respectively; **mode3/mode4** to reflect 0: Rectangular, 1: Triangular, and 2: Exponential. Find the Fourier transform (FT) of these two signals. Set up another case structure by using **mode1** `switch_expression` to control the combination method of x1 and x2 in the time domain (0: Linear combination, 1: Convolution, 2: Multiplication). Finally, set up a case structure by using **mode2** `switch_expression` to control the combination method in the frequency domain. This operation is performed on the FT transform of x1 and x2. First, set **Pulse width**, **Time shift**, and **Time scale** as control parameters. **Pulse width** controls the number of ones in the pulse, which is used to increase or decrease the pulse width. **Time shift** adds zeros before the pulse to provide a time delay. **Time scale** is multiplied with the time increment dt to ensure appropriate scaling of the pulse. The MATLAB function `fft` can be used to determine the FT of the continuous signal. Combine the signals in the frequency domain and set the combination method (linear combination, convolution or multiplication) via the parameter **mode2**. Compute the FT of the time domain combinations and the inverse FT of the frequency domain combinations using the functions `fft` and `ifft`. To shift the zero-frequency component to the center of the spectrum, use the MATLAB function `fftshift`. Note that it is not necessary to use fftshift. However, its use allows a better representation of signals in the frequency domain. Finally, determine the magnitude and phase of the FT using the functions `abs` and `angle`, respectively. Display the input signals and their combinations using the `plot` function. Save the MATLAB script using the name L6_1. The MATLAB code for this example is shown in Figure 6.1 in two pages.

Next, open a new script and write a verification code to evaluate the function L6_1 and save it as L6_1_testbench. Evaluate the output for different control parameters. See Figure 6.2. After running L6_1_testbench, follow the steps outlined in L4_1 to generate the C code and then place it into the shell provided. If using MATLAB R2016b or later, it is required to add a specific header file into the *jni* folder separately. This header file (`tmwtypes.h`) can be found in the MATLAB root with this path `MATLAB root\R2019b\extern\include`. Figures 6.3 and 6.4 show the app screen on an Android smartphone as well as the setting parameters.

```
1    function [x1,x2,x,x3,y1_mag,y1_phase,y2_mag,y2_phase,y_mag,y_phase,y3_mag,y3_phase,f]=...
2       L6_1(dt,del1,d1,a,del2,d2,a1,a2,mode1,mode2,mode3,mode4)
3
4 -  fs=1/dt;
5 -  tau = a*dt;
6 -  T=1;
7 -  N=floor(T/tau);
8 -  n1=floor(del1/dt);
9 -  n2=floor(del2/dt);
10 - a11=zeros(1,floor(d1/dt));
11 - a12=zeros(1,abs(N-n1));
12 - a21=zeros(1,floor(d2/dt));
13 - a22=zeros(1,abs(N-n2));
14
15   % Finding the next power of 2 for N because MATLAB coder only decodes the
16   % power of 2 for FFT
17 - N1 = max(2^(ceil(log2(N+d1/dt))),2^(ceil(log2(N+d2/dt))));
18
19   % Generating signal 1
20 - x1 = zeros(1,N1);
21 - x2 = zeros(1,N1);
22 - switch mode3
23 - case 0
24 - x1=[a11 ones(1,n1) a12];
25 - case 1
26 - x1=[a11 (1:n1)./n1 a12];
27 - case 2
28 - x1=[a11 exp(-(1:n1)/n1) a12];
29 - end
30
31   % Generating signal 2
32 - switch mode4
33 - case 0
34 - x2=[a21 ones(1,n2) a22];
35 - case 1
36 - x2=[a21 (1:n2)./n2 a22];
37 - case 2
38 - x2=[a21 exp(-(1:n2)/n2) a22];
39 - end
```

Figure 6.1: L6_1 CTFT and its properties. (*Continues.*)

```
40
41        % Zero padding
42 -      x1=[x1 zeros(1,(N1-length(x1)))];
43 -      x2=[x2 zeros(1,(N1-length(x2)))];
44
45        % Operation in time domain
46 -      x=zeros(1,N1);
47 -      switch mode1
48 -      case 0
49 -      x=a1*x1+a2*x2;
50 -      case 1
51 -      x=conv(x1,x2);
52 -      x=x(1:N1);
53 -      case 2
54 -      x=x1.*x2;
55 -      end
56
57        % Taking FFT
58 -      y1=1/N1*fftshift(fft(x1));
59 -      y2=1/N1*fftshift(fft(x2));
60 -      y=1/N1*fftshift(fft(x));
61
62        % Operation in frequency domain
63 -      y3=zeros(1,N1);
64 -      switch mode2
65 -      case 0
66 -      y3=a1*y1 + a2*y2;
67 -      case 1
68 -      y3=conv(y1,y2);
69 -      y3=y3(N1/2+1:3*N1/2);
70 -      case 2
71 -      y3=N1*y1.*y2;
74        % Taking IFFT and plotting
75 -      x3=abs(N1*ifft(fftshift(y3)));
76 -      y1_mag=abs(y1);
77 -      y1_phase=angle(y1);
78 -      y2_mag=abs(y2);
79 -      y2_phase=angle(y2);
80 -      y3_mag=abs(y3);
81 -      y3_phase=angle(y3);
82 -      y_mag=abs(y);
83 -      y_phase=angle(y);
84 -      f=1/a*(-fs/2:fs/length(y1):fs/2);
```

L6_1   Ln 69   Col 22

Figure 6.1: (*Continued.*) L6_1 CTFT and its properties.

```
L6_1_testbench.m  ✕  +
1 -   clc; clear; close all;
2 -   dt=0.001; %Time interval between samples
3 -   del1=.6; % Pulse width of first input
4 -   d1=0.2; % Time shift of first input
5 -   a=2; % Time scale
6 -   del2=0.3; % Pulse width of second input
7 -   d2=0.6; % Time shift of second input
8 -   a1=1; % Scale1 for linear combination
9 -   a2=3; % Scale2 for linear combination
10 -  mode1=1; %Operation in time domain:: 0: Linear combination, 1: Convolution, 2: Multiplication
11 -  mode2=2; %Operation in frequency domain:: 0: Linear combination, 1: Convolution, 2: Multiplication
12 -  mode3=1; %Signal 1:: 0:Rectangular, 1: Triangular, 2: Exponential
13 -  mode4=1; %Signal 2:: 0:Rectangular, 1: Triangular, 2: Exponential
14 -  [x1,x2,x,x3,y1_mag,y1_phase,y2_mag,y2_phase,y_mag,y_phase,y3_mag,y3_phase,f]=...
15        L6_1(dt,del1,d1,a,del2,d2,a1,a2,mode1,mode2,mode3,mode4);
16    %% Verification of the Results
17 -  figure(1)
18 -  plot(dt*(1:length(x1)),x1);grid on;
19 -  hold all
20 -  plot(x1,'r')
21 -  xlabel('t')
22 -  ylabel('x1')
23 -  figure(2)
24 -  plot(dt*(1:length(x2)),x2);grid on;
25 -  xlabel('t')
26 -  ylabel('x2')
27 -  figure(3)
28 -  plot(dt*(1:length(x3)),x3);grid on;
29 -  xlabel('t')
30 -  title('x3:IFFT of the combination in the frequency domain')
31 -  figure(4)
32 -  plot(dt*(1:length(x)),x);grid on;
33 -  xlabel('t')
34 -  title('x: combined signals in time domain')
                                                                          script                    Ln 2 Col 1
```

Figure 6.2: **L6_1_testbench** script for verifying CTFT properties.

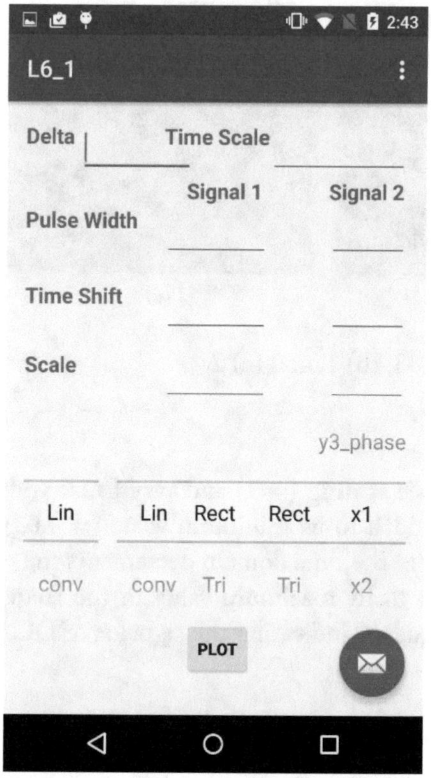

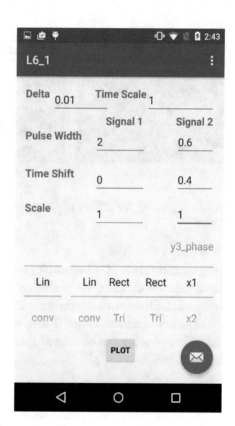

Figure 6.3: Smartphone app screen.

Figure 6.4: Parameter settings.

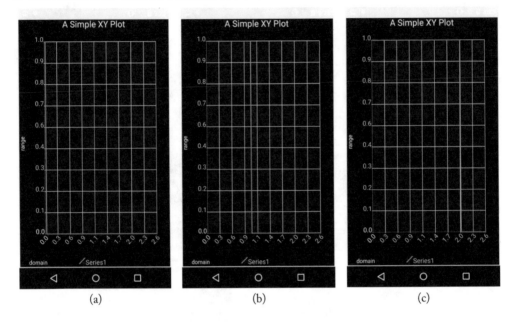

Figure 6.5: Signal x1 for different pulse widths: (a) 0.03, (b) 1, and (c) 2.

L6.1.1 Varying Pulse Width

Keep the default values of **Time shift** ($= 0$) and **Time scaling** ($= 1$) and vary **Pulse width** of the rectangular pulse. First, set the value of **Pulse width** to its minimum value ($= 0.03$) and then increase it. Observe that increasing **Pulse width** in the time domain decrements the width in the frequency domain. When **Pulse width** is set to its maximum value in the frequency domain, only one value can be seen at the center frequency indicating the signal is of DC type, see Figures 6.5–6.7.

L6.1.2 Time Shift

Next, for a fixed **pulse width**, vary **time shift**. Observe that the phase spectrum changes but the magnitude spectrum remains the same. If the signal $f(t)$ is shifted by a constant t_0, its FT magnitude does not change, but the term $-\omega t_0$ gets added to its phase angle, see Figures 6.8–6.10.

L6.1.3 Time Scaling

Observe that increasing the parameter **Time scaling** makes the spectrum wider. This indicates that compressing the signal in the time domain leads to expansion in the frequency domain; see Figures 6.11–6.13.

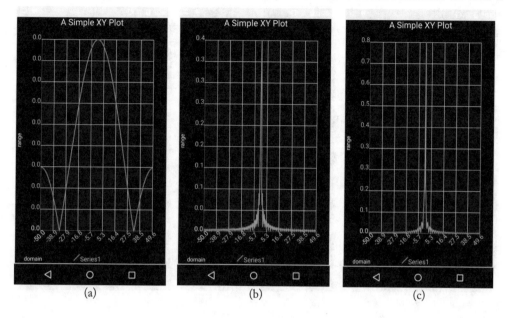

Figure 6.6: Magnitude spectrum of signal x1 for different pulse widths: (a) 0.03, (b) 1, and (c) 2.

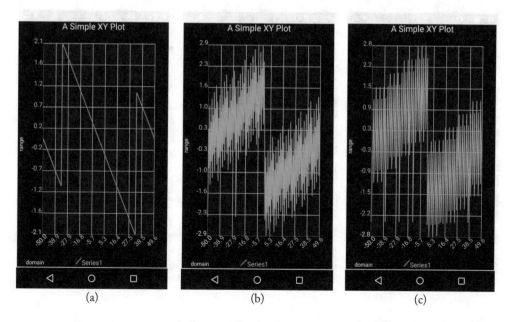

Figure 6.7: Phase of signal x1 for different pulse widths: (a) 0.01, (b) 1, and (c) 2.

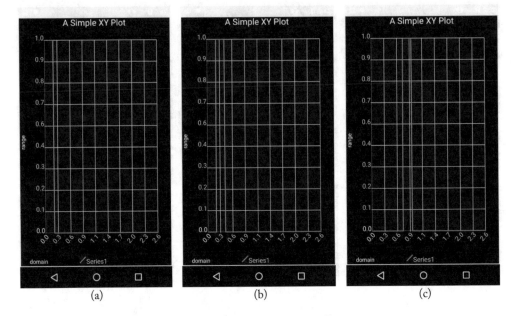

Figure 6.8: Signal x1 for different time shifts: (a) 0, (b) 0.2, and (c) 0.7.

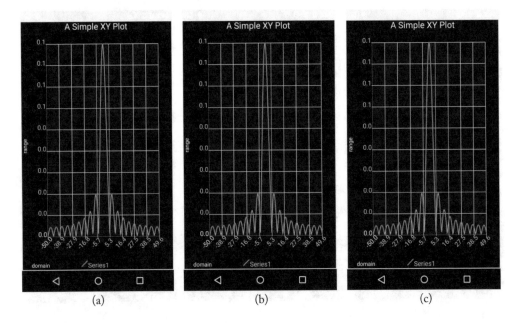

Figure 6.9: Magnitude spectrum of signal x1 for different time shifts: (a) 0, (b) 0.2, and (c) 0.7.

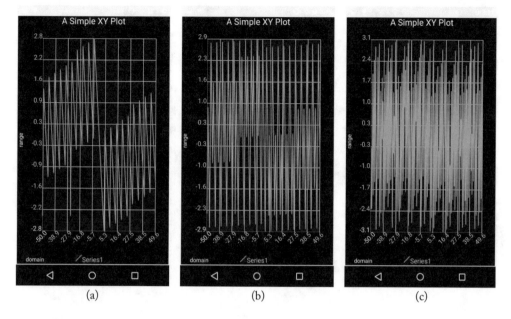

Figure 6.10: Phase of signal x1 for different time shifts: (a) 0, (b) 0.2, and (c) 0.7.

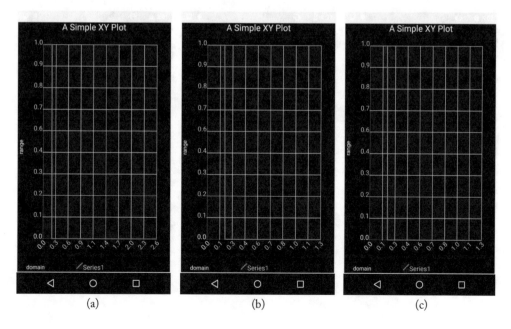

Figure 6.11: Signal x1 for different time scalings: (a) 1, (b) 2, and (c) 3.

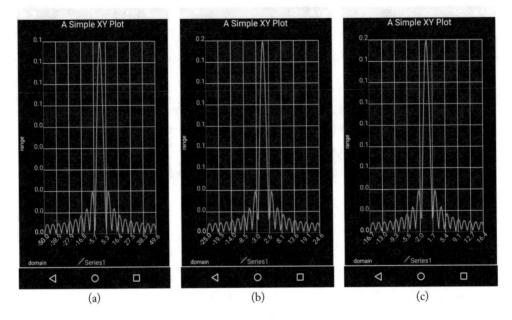

Figure 6.12: Magnitude spectrum of signal x1 for different time scalings: (a) 1, (b) 2, and (c) 3.

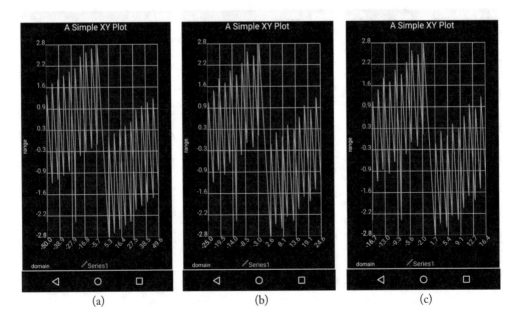

Figure 6.13: Phase of signal x1 for different time scalings: (a) 1, (b) 2, and (c) 3.

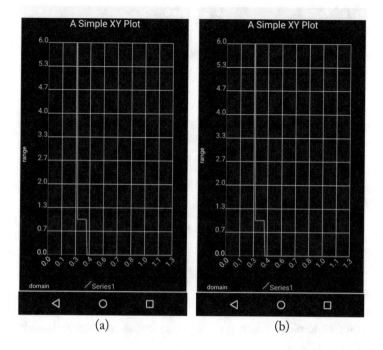

Figure 6.14: (a) Plot of $a_1x_1(t) + a_2x_2(t)$ and (b) plot of the inverse FT of $a_1X_1(\omega) + a_2X_2(\omega)$.

L6.1.4 Linearity

In this section, two signals are combined to examine the linearity property of FT. Select **Linear Combination** for the **Time domain** and **Frequency domain** combination method. This selection combines the two time signals $x_1(t)$ and $x_2(t)$ linearly with the scaling factors a_1 and a_2, producing a new signal $a_1x_1(t) + a_2x_2(t)$. Figure 6.14a displays the FT of this linear combination. The linear combination in the frequency domain produces a new signal $a_1X_1(\omega) + a_2X_2(\omega)$. Figure 6.14b also displays the inverse FT of this combination. Observe that both combinations produce the same result in the time and frequency domains as shown in Figures 6.14 and 6.15, respectively. For this example, a_1 and a_2 are set to 1 and 5, respectively. The pulse width of $x_1(t)$ and $x_2(t)$ are set to 0.4 and 0.3, respectively.

L6.1.5 Time Convolution

In this part, two signals are convolved in the time domain to examine the time-convolution property of FT. Select **Convolution** for **Time domain** and **Multiplication** for **Frequency domain**. This selection produces and displays a new signal, $x_1(t) * x_2(t)$ by convolving the two time signals $x_1(t)$ and $x_2(t)$. Multiplication in the frequency domain produces a new signal $X_1(\omega)X_2(\omega)$. The inverse FT of this multiplied signal is displayed in Figure 6.18a. Note that

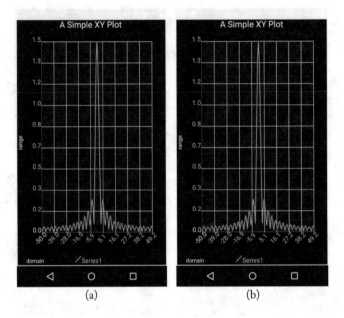

Figure 6.15: (a) Magnitude plot of $a_1 x_1(t) + a_2 x_2(t)$ in the frequency domain and (b) magnitude plot of $a_1 X_1(\omega) + a_2 X_2(\omega)$.

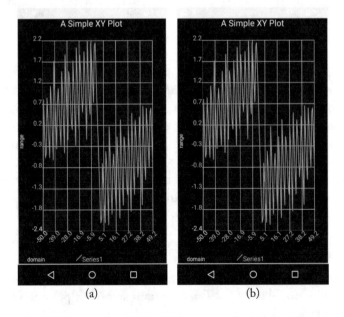

Figure 6.16: (a) Phase plot of $a_1 x_1(t) + a_2 x_2(t)$ in the frequency domain and (b) phase plot of $a_1 X_1(\omega) + a_2 X_2(\omega)$.

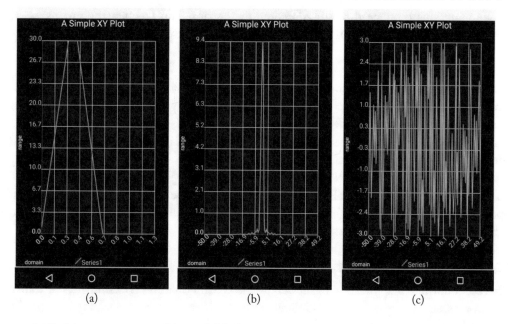

Figure 6.17: (a) $x_1(t) * x_2(t)$, (b), and (c) Fourier transform magnitude and phase of $x_1(t) * x_2(t)$, respectively.

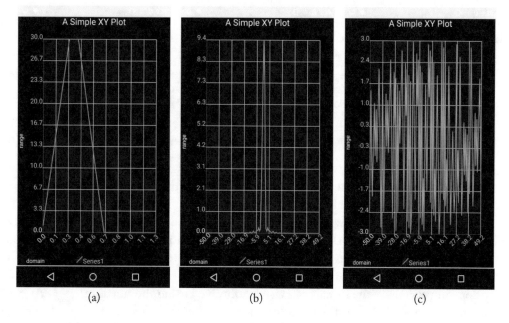

Figure 6.18: (a) Inverse of $X_1(\omega)X_2(\omega)$, (b), and (c) magnitude and phase of $X_1(\omega)X_2(\omega)$, respectively.

both combinations produce the same outcome in the time and frequency domains, see Figures 6.17 and 6.18. This verifies the frequency-convolution property stated in Table 6.1.

L6.1.6 Frequency Convolution

In this section, two signals are convolved in the frequency domain to examine the frequency-convolution property of FT. Select **Convolution** for **Frequency domain** and **Multiplication** for **Time domain**. This selection convolves the two-time signals $X_1(\omega)$ and $X_2(\omega)$ to produce a new signal $X_1(\omega) * X_2(\omega)$. The inverse FT of the convolved signal is displayed in Figure 6.20a. **Multiplication** in **Time domain** produces a new signal $x_1(t)x_2(t)$. The FT of this multiplied signal is also displayed in Figure 6.19. Note that both combinations produce the same outcome in the time and frequency domains; see Figures 6.19 and 6.20. This verifies the frequency-convolution property stated in Table 6.1. In this example, the triangular and exponential functions are used for the first and second inputs, respectively.

L6.2 AMPLITUDE MODULATION

In this section, the FT application of modulation and demodulation is examined. For transmission purposes, signals are normally modulated with a high-frequency carrier. A typical amplitude modulated signal can be written as

$$x(t) = x_m(t) \cos (2\pi f_c t), \qquad (6.6)$$

where $x_m(t)$ is called the message waveform, which contains the data of interest, and f_c is the carrier wave frequency. Using the identity

$$\cos (2\pi f_c t) = \frac{1}{2} \left(e^{2\pi f_c t} + e^{-2\pi f_c t} \right) = \frac{1}{2} \left(e^{\omega_c t} + e^{-\omega_c t} \right) \qquad (6.7)$$

and the frequency shift property of CTFT, the CTFT of $x(t)$ can easily be derived to be

$$X(\omega) = \frac{1}{2} \left(X_m(\omega - \omega_c) + X_m(\omega + \omega_c) \right). \qquad (6.8)$$

At the receiver, some noisy version of this transmitted signal is received. The signal information resides in the envelope of the modulated signal, and thus an envelope detector can be used to recover the message signal.

Figures 6.21 and 6.22 show the MATLAB function and the test bench code for the demodulation system. The MATLAB function is named L6_2 and the test bench L6_2_testbench. In this example, the combination of two sine waves are used to serve as a message signal. The signal is modulated with a high-frequency carrier, and some random noise is added. The frequency domain versions of the signals can also be observed using the function `fft`. As stated in Equation (6.8), the CTFT of the modulated signal is merely a frequency-shifted version of the original signal. In single sideband (SSB) modulation, only one side of the spectrum is transmitted due to symmetry. That is, just one side of the spectrum is taken and converted into a

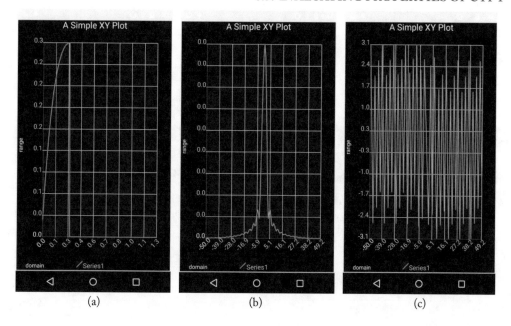

Figure 6.19: (a) $x_1(t) \times x_2(t)$, (b), and (c) Fourier transform magnitude, and phase of $x_1(t) \times x_2(t)$, respectively.

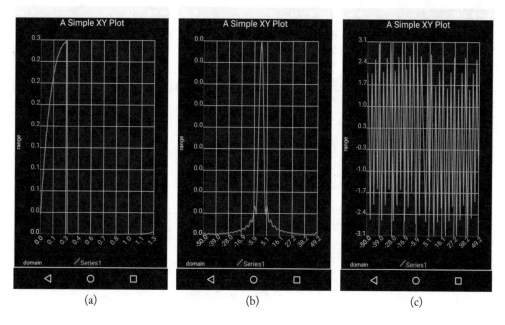

Figure 6.20: (a) Inverse of $X_1(\omega) * X_2(\omega)$, (b), and (c) magnitude, and phase of $X_1(\omega) * X_2(\omega)$, respectively.

```matlab
Editor - C:\Users\Axa180003\Desktop\MATLAB Examples\Chapter 6\L6_2\L6_2.m
L6_2.m  ×  +
1      function [z1,z2,z3,z4,y1,y2,y3,y4]=L6_2(fc)
2      fs=5*fc; % fc=3000;
3      dt=1/fs;
4      N1=2^(ceil(log2(0.1*fs))); %Since FFT function requires the input to be
5      T=dt*N1;                   %a power of 2, obtaining the next power of 2
6      t1=0:1/fs:T;
7      %% Generating the signal
8      f0 = 31/150*fc;
9      f1 = 1/30*fc;
10     z1=sin(2*pi*f0*t1)+sin(2*pi*f1*t1);
11     %% Amplitude modulating the signal
12     z2=z1.*sin(2*pi*fc*t1);
13     %% Adding noise
14     z3=z2;
15     %sprev = rng(11235813213455, 'twister');
16     z3 = z3 + 0.1*rand(1,length(z2));
17     n=length(z1);
18     j=complex(0,1);
19     y1=abs(1/n*fft(z1,N1)); % FFT of signal
20     y2=abs(1/n*fft(z2,N1)); % FFT of modulated signal
21     y3=(1/n*fft(z3,N1));    % FFT of noisy signal
22     y4 = zeros(1,N1) + j*zeros(1,N1);
23     %% Retreiving the part of the spectrum that contains all signal and no noise
24     y4(1:256) = y3(round(fc*T)+1:round(fc*T)+256);
25     z4=real(n*ifft(y4)); % Taking IFFT to obtain the original signal
26     y1=(fftshift(y1));
27     y2=(fftshift(y2));
28     y3=abs(fftshift(y3));
29     y4=abs(fftshift(y4));
30     %f=-fs/2:fs/length(y1):fs/2;
31
```

Figure 6.21: **L6.2** amplitude modulation function.

Figure 6.22: L6.2_testbench of amplitude modulation function.

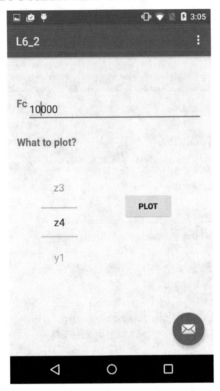

Figure 6.23: L6_2 initial screen on smartphone.

time signal using the function `ifft`. Run L6_2_testbench and then generate the C code and incorporate it in the shell provided. Figure 6.23 shows the initial screen on a smartphone platform. If using MATLAB R2016b or later, it is required to add a specific header file into the *jni* folder, separately. This header file (`tmwtypes.h`) can be found in the MATLAB root with this path `MATLAB root\R2019b\extern\include` . The frequency f_c needs to be specified. The desired output for plotting can be selected from the **What to plot** option, z1, z2, z3, and z4, corresponding to Message signal, Modulated signal, Received signal (noise added), and Demodulated signal, respectively, where y1, y2, y3, and y4 correspond to the spectrum magnitude of these signals. Figure 6.24 shows the Message signal, Modulated signal, Received signal (noise added), and Demodulated signal; Figure 6.25 illustrates the magnitude of these signals.

L6.3 NOISE REDUCTION

When a signal passes through a channel, it normally gets corrupted by channel noise. Various electronic components used in a transmitter or receiver may also cause additional noise. Noise

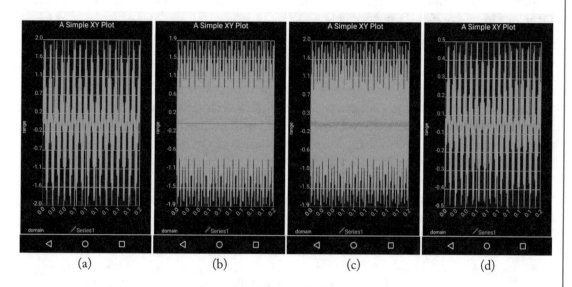

Figure 6.24: (a) Message signal, (b) Modulated signal, (c) Received signal (noise added), and (d) Demodulated signal.

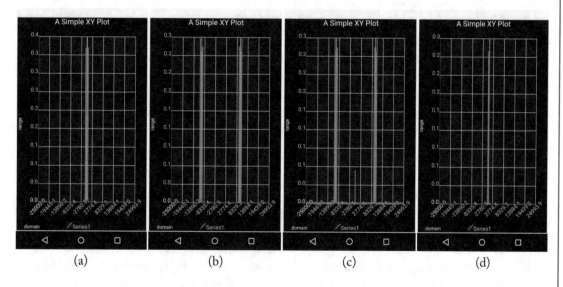

Figure 6.25: Frequency domain: (a) Message signal, (b) Modulated signal, (c) Received signal (noise added), and (d) Demodulated signal.

reduction is an essential module of any signal processing system. This section presents a simple technique to reduce high-frequency noise.

Open a *NewScript* and write a MATLAB code for a noise reduction system. Consider a sine wave signal at 2 Hz frequency. Add some high-frequency noise to this signal and then remove the high-frequency components in the frequency domain. Finally, move the signal back into the time domain using the inverse FT. Use two control parameters named **Time frame width** and **Frame number** to extract a segment of the speech signal before computing Fourier transform. Add together three sine and cosine waves with frequencies of 3.5, 3, and 2.8 kHz to create a high-frequency noise. Then, add a scaled version of the noise signal to the signal with the **Noise Level parameter** set as a control parameter. Compute the FT of the **Noise added signal** using the function `fft` .

To remove the high-frequency noise components, use a simple lowpass filter by removing the frequency components over a certain threshold (50%, for example). After removing the high-frequency components, transform the signal back into the time domain using the function `ifft` . To get a display of the absolute and centered frequency spectrum, use the functions *abs* and `fftshift` .

Figures 6.26 and 6.27 show the MATLAB function for the noise reduction and L6_3_testbench, respectively. Name the MATLAB function L6_3 and the test script L6_3_testbench. First run L6_3_testbench and observe the results for different parameters. Then, generate the C code using the MATLAB Coder and incorporate it into the shell provided by following the steps mentioned in Chapter 3. If using MATLAB R2016b or later, it is required to add a specific header file into the *jni* folder, separately. This header file (`tmwtypes.h`) can be found in the MATLAB root with this path `MATLAB root\R2019b\extern\include` . Figure 6.28 shows the screen of the app on an Android smartphone with some sample setting parameters. In the main screen, one can see there is a plot picker **What to plot**? z1, z2, and z3 which can be selected denoting the segment of the original signal, the signal with high frequency noise, and the noise reduced signal, respectively. y1, y2, and y3 denote the magnitude responses of z1, z2, and z3, respectively. In Figures 6.29 and 6.30, the plots are shown.

6.4 LAB EXERCISES

6.4.1 CIRCUIT ANALYSIS

Find and plot the frequency response (both magnitude and phase spectrum) of each of the circuits shown in Figure 6.31. Set the values of R, L, and C as control parameters.

Consider a message containing some hidden information. Furthermore, to make it interesting, suppose the message contains a name. Assume that the message was coded using the amplitude modulation scheme as follows [4]:

$$x(t) = x_{m1}(t) \cos(2\pi f_1 t) + x_{m2}(t) \cos(2\pi f_2 t) + x_{m3}(t) \cos(2\pi f_3 t), \qquad (6.9)$$

Editor - C:\Users\Axa180003\Desktop\MATLAB Examples\Chapter 6\L6_3\L6_3.m

L6_3.m

```
function [z1,z2,z3,y1,y2,y3] = L6_3(fs, a, fw, fn)
%Plotting the signal
dt=1/fs;
t1=0:dt:5;
z=sin(2*pi*2*t1);
z1 = z((fn-1)*floor(fw*fs)+1:fn*floor(fw*fs));
n = length(z1);
% m = [zeros(1,(fn-1)*n) ones(1,n) zeros(1,n)];
N1 = 2^(ceil(log2(n))); %Since FFT function requires the input to be a
%power of 2, obtaining the next power of 2
%% Adding noise to the input
t=0:dt:(n-1)*dt;
f1=3500;
f2=3000;
f3=2800;
noise = sin(2*pi*f1*t) + cos(2*pi*f2*t) + cos(2*pi*f3*t);
z2 = z1+a*noise;
%% Zero padding so that z1 and z2 are length of power of 2
z1 = [z1 zeros(1,N1-length(z1))];
z2 = [z2 zeros(1,N1-length(z2))];
y1=fft(z1); % Taking FFT
y2=fft(z2); % Taking FFT
y3=y2;
y3(floor(N1/8):N1-floor(N1/8))=0; % Removing high frequency noise component
z3=real(ifft(y3));
y1=1/n*abs(fftshift(y1)); % Taking the inverse of all FFTs
y2=1/n*abs(fftshift(y2));
y3=1/n*abs(fftshift(y3));
z1=z1(1:n);
z2=z2(1:n);
z3=z3(1:n);
```

Figure 6.26: L6_3 noise reduction function.

```
Editor - C:\Users\Axa180003\Desktop\MATLAB Examples\Chapter 6\L6_3\L6_3_testbench.m          ⊙ ×

  L6_3_testbench.m  ×  +

 1 -   clc;clear;close all;
 2
 3 -   fs=5000; %Sampling frequency
 4 -   a=0.1; % Noise level
 5 -   fw=0.1; % Frame width in sec
 6 -   fn=10; % Frame number
 7
 8 -   [z1,z2,z3,y1,y2,y3] = L6_3(fs, a, fw, fn);
 9     %% Verification of Results
10
11 -   figure(1);
12 -   subplot(1,3,1)
13 -   plot(z1)
14 -   title('Segment of the original signal')
15 -   subplot(1,3,2)
16 -   plot(z2)
17 -   title('Signal with high frequency noise')
18 -   subplot(1,3,3)
19 -   plot(z3)
20 -   title('Noise reduced')
21 -   figure(2);
22 -   subplot(1,3,1)
23 -   plot(y1)
24 -   title('frequency domain: Segment of the original signal')
25 -   subplot(1,3,2)
26 -   plot(y2)
27 -   title('frequency domain: Signal with high frequency noise')
28 -   subplot(1,3,3)
29 -   plot(y3)
30 -   title('frequency domain: Noise reduced')
```

Figure 6.27: L6_3_testbench of noise reduction function.

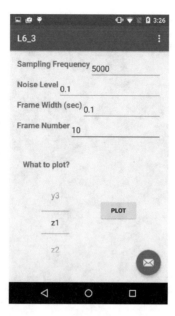

Figure 6.28: Noise reduction smartphone app.

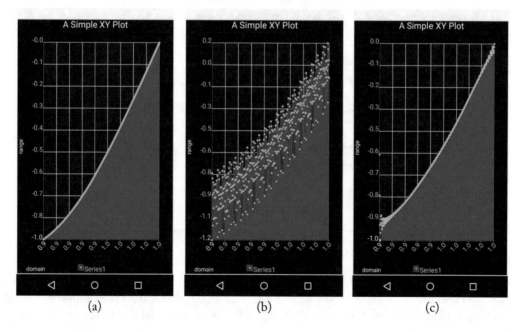

Figure 6.29: (a) Plot of segment of the original sine wave (z1), (b) the noisy signal (z2), and (c) the noise reduced signal (z3).

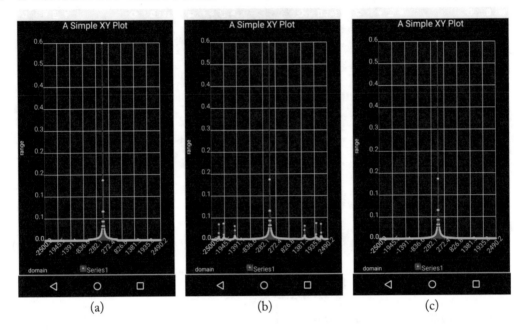

Figure 6.30: (a) Plot of the magnitude of the original sine wave (y1), (b) the noisy signal (y2), and (c) the noise reduced signal (y3) in the frequency domain.

Table 6.2: Alphabet letters encoded with Morse

A	. −	H	....	O	− − −	V	... −
B	− ...	I	..	P	. − − .	W	. − −
C	− . − .	J	. − − −	Q	− − . −	X	− .. −
D	− ..	K	− . −	R	. − .	Y	− . − −
E	.	L	. − ..	S	...	Z	− − ..
F	.. − .	M	− −	T	−		
G	− − .	N	− .	U	.. −		

where $x_{m1}(t)$, $x_{m2}(t)$, and $x_{m3}(t)$ are the (message) signals containing the three letters of the name. More specifically, let each of the signals $x_{m1}(t)$, $x_{m2}(t)$, and $x_{m3}(t)$ correspond to a single letter of the alphabet. Encode these letters using the International Morse Code as indicated in Table 6.2 [4]:

As shown in Table 6.2, to encode the letter A, one needs only a dot followed by a dash. That is, only two prototype signals are needed—one to represent the dash and one to represent

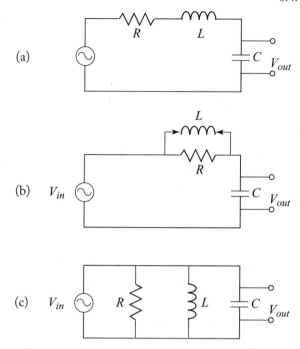

Figure 6.31: Linear RLC circuits.

the dot. Thus, for instance, to represent the letter A, set $x_{m1}(t) = d(t) + dash(t)$, where $d(t)$ represents the dot signal and *dash(t)* the dash signal. Similarly, to represent the letter O, set $x_{m1(t)=3}dash(t)$.

Find the prototype signals $d(t)$ and *dash(t)* in the supplied file *morse.mat*. After loading the file *morse.mat*:

```
>>load morse
```

The signals $d(t)$ and *dash(t)* can be located in the vectors dot and dash, respectively. The hidden signal, which is encoded per Equation (6.9), contains the letters of the name in the vector xt. Let the three modulation frequencies f_1, f_2, and f_3 be 20, 40, and 80 Hz, respectively.

1. Using the amplitude modulation property of the CTFT, determine the three possible letters and the hidden name. (Hint: Plot the CTFT of xt. Use the values of T and τ contained in the file.)

2. Explain the strategy used to decode the message. Is the coding technique ambiguous? That is, is there a one-to-one mapping between the message waveforms $(x_{m1}(t), x_{m2}(t), x_{m3}(t))$

and the alphabet letters? Or can you find multiple letters that correspond to the same message waveform?

6.4.2   DOPPLER EFFECT

The Doppler Effect phenomenon was covered in the previous chapter. In this exercise, let us examine the Doppler Effect with a real sound wave rather than a periodic signal. The wave file *firetrucksiren.wav* provided in the book software package contains a firetruck siren. Read the file using the MATLAB function `audioread` and produce its upscale and downscale versions. Show the waves in the time and frequency domains (find the CTFT). Figure 6.32 shows the original sound, the sound as the vehicle approaches and the sound after the vehicle passes by in both the time-domain and frequency domains.

6.4.3   DIFFRACTION OF LIGHT

The diffraction of light can be described via a Fourier transform [5]. Consider an opaque screen with a small slit being illuminated by a normally incident uniform light wave, as shown in Figure 6.33.

Considering that $d \gg \pi l_1^2 / \lambda$ provides a good approximation for any l_1 in the slit, the electric field strength of the light striking the viewing screen can be expressed as [5]

$$E_0(l_0) = K \frac{e^{j(2\pi d/\lambda)}}{j\lambda d} e^{j(\pi/\lambda d)l_0^2} \int_{-\infty}^{\infty} E_1(l_1) e^{-j(2\pi/\lambda d)l_0 l_1} \, dl_1, \tag{6.10}$$

where

E_1 = field strength at diffraction screen
E_0 = field strength at viewing screen
K = constant of proportionality
λ = wavelength of light

The above integral is in fact Fourier transformation in a different notation. One can write the field strength at the viewing screen as [5]

$$E_0(l_0) = K \frac{e^{j(2\pi d/\lambda)}}{j\lambda d} e^{j(\pi/\lambda d)l_0^2} CTFT \{E_1(t)\}_{f \to l_0/\lambda d}. \tag{6.11}$$

The intensity $I(l_0)$ of the light at the viewing screen is the square of the magnitude of the field strength. That is,

$$I(l_0) = |E_0(l_0)|^2. \tag{6.12}$$

Plot the intensity of the light at the viewing screen. Set the slit width to this range (0.5–5 mm), the wavelength of light λ to this range (300–800 nm), and the distance of the viewing

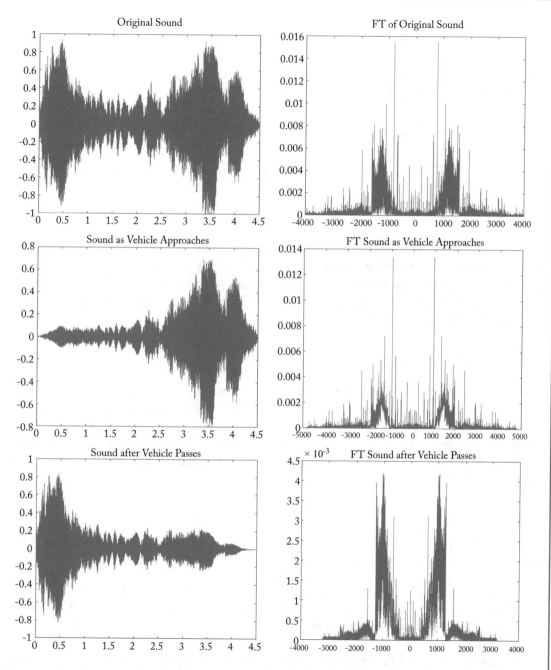

Figure 6.32: Doppler Effect system, original sound, sound as vehicle approaches, and sound after vehicle passes in both time-domain and frequency domain.

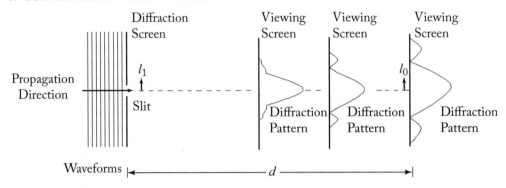

Figure 6.33: Diffraction of light.

screen d to this range (10–200 m) as control parameters. Assume the constant of proportionality is 10^{-3}, and the electric field strength at the diffraction screen is 1 V/m.

Now replace the slit with two slits, each 0.1 mm in width, separated by 1 mm (center-to-center) and centered on the optical axis. Plot the intensity of light in the viewing screen by setting the parameters as controls.

6.5 REAL-TIME LABS

LR6_2 – This real-time lab is a noise reduction application. The input signal is modulated with a tone at fs/5 then noise is added to it. One should be able to choose the output to plot. The choices are z2, the modulated input signal z3, the modulated input signal with noise y1, the fft of the original signal y2, the fft of the modulated signal y3, the fft of the modulated signal with noise y4, the noise removed version of y3 and y4, the noise removed version of y3 and z4, and the inverse fft of the noise removed signal.

ANDROID STEPS

These steps modify the LR5_2 lab. Copy the LR5_2 directory to LR6_2 and rename the LR5_2 files to LR6_2. Update the project files before updating the codes.

1. Update the applicationId in app/build.gradle to have LR6_2 in the name instead of LR5_2. Remove the old C code from app/src/main/jni and copy the new MATLAB generated code into the same directory.

2. If using MATLAB R2016b or later, it is required to add a specific header file into the *jni* folder separately. This header file (`tmwtypes.h`) can be found in the MATLAB root with the following path:

```
MATLAB root\R2019b\extern\include
```

3. Update the MATLAB generated files in the file CMakeLists.txt. The section in the file should look like the following block of code. The order of the files does not matter.

```
add_library( # Sets the name of the library.
             matlabNative

             # Sets the library as a shared library.
             SHARED

             # Provides a relative path to your source file(s).
             src/main/jni/eml_rand_mt19937ar_stateful.c
             src/main/jni/eml_rand_mt19937ar_stateful.h
             src/main/jni/fft.c
             src/main/jni/fft1.c
             src/main/jni/fft1.h
             src/main/jni/fft.h
             src/main/jni/fftshift.c
             src/main/jni/fftshift.h
             src/main/jni/LR6_2.c
             src/main/jni/LR6_2.h
             src/main/jni/LR6_2_data.c
             src/main/jni/LR6_2_data.h
             src/main/jni/LR6_2_emxAPI.c
             src/main/jni/LR6_2_emxAPI.h
             src/main/jni/LR6_2_emxutil.c
             src/main/jni/LR6_2_emxutil.h
             src/main/jni/LR6_2_initialize.c
             src/main/jni/LR6_2_initialize.h
             src/main/jni/LR6_2_terminate.c
             src/main/jni/LR6_2_terminate.h
             src/main/jni/LR6_2_types.h
             src/main/jni/MatlabNative.c
             src/main/jni/rand.c
             src/main/jni/rand.h
             src/main/jni/rtGetInf.c
```

```
        src/main/jni/rtGetInf.h
        src/main/jni/rtGetNaN.c
        src/main/jni/rtGetNaN.h
        src/main/jni/rtwtypes.h
        src/main/jni/rt_nonfinite.c
        src/main/jni/rt_nonfinite.h
        src/main/jni/tmwtypes.h
    )
```

4. Moving onto *app/src/main/AndroidManifest.xml*, update the LR5_2 names to LR6_2.

5. The file *app/src/main/res/xml/prefs.xml* will need to be edited to remove the resistance, capacitance, frequency, and output settings similar to the previous labs. This time a List-Preference of the output choice is made:

```
<ListPreference
        android:key="coice1"
        android:title="What to plot?"
        android:summary="Default: z2"
        android:defaultValue="0"
        android:entries="@array/choiceOptions"
        android:entryValues="@array/choiceValues" />
```

Since the array references, `@array/choiceOptions` and `@array/choiceValues`, are used, they need to get added to the file *app/src/main/res/values/arrays.xml*. Remove `systemOptions`, `systemValues`, `choiceOptions`, and `choiceValues`.

```
<string-array name="choiceOptions">
    <item>z2</item>
    <item>z3</item>
    <item>y1</item>
    <item>y2</item>
    <item>y3</item>
    <item>y4</item>
    <item>z4</item>
</string-array>
<string-array name="choiceValues">
```

```
    <item>0</item>
    <item>1</item>
    <item>2</item>
    <item>3</item>
    <item>4</item>
    <item>5</item>
    <item>6</item>
</string-array>
```

6. Moving on to the code changes, update the file app/src/main/java/com/dsp/matlab/Filter.java for the function calls by adding resistance, capacitance, frequency, and display choice,

```
currentFrame.setFiltered(Filters.compute(currentFrame.getFloats(),
Settings.Fs, Settings.choice, Settings.blockSize));
```

7. Now, in the file app/src/main/java/com/dsp/matlab/Filters.java, update the compute method as follows:

```
public static native float[] compute( float[] in, int Fs,
int choice, int frameSize );
```

8. To make sure the coefficient input is applied to the settings, update the `updateSettings()` method in app/src/main/java/com/dsp/matlab/RealTime.java and add the following line within. The previous settings for resistance, capacitance, coefficients, frequency, and system can be removed.

```
Settings.setChoice(Integer.parseInt(preferences.getString
("choice1", "0")));
```

9. Update app/src/main/java/com/dsp/matlab/Settings.java to store and retrieve the settings added. Again the resistance, capacitance, coefficients, frequency, and system-related variables and methods can be removed.

```
public static float choice = 0;
public static void setChoice(int choice1){choice=choice1;}
```

10. Finally, update the app/src/main/jni/MatlabNative.c interface file. This file is the interface from Java to native C. The entire file is listed below.

```
#include <jni.h>
#include <stdio.h>
#include "rt_nonfinite.h"
#include "LR6_2.h"
#include "LR6_2_terminate.h"
#include "LR6_2_emxAPI.h"
#include "LR6_2_intialize.h"

jfloatArray
Java_com_dsp_matlab_Filters_compute(JNIEnv *env, jobject this,
jfloatArray, input, jint Fs, jint choice, jint frameSize)
{
    jfloatArray output = (*env)->NewFloatArray(env, frameSize);
    float *_in = (*env)->GetFloatArrayElements(env, input, NULL );

    //compute
    emxArray_real32_T *result;
    int X_size[2] = {1, frameSize};
    emxInitArray_real32_T(&result,2);

    LR6_2(_in, X_size,Fs, choice, result);

    (*env)->SetFloatArrayRegion(env,output,0,frameSize,
                                result->data);
    emxDestroyArray_real32_T(result);
    (*env)->ReleaseFloatArrayElements(env, input, _in, 0);
    return output;
}

Void Java_com_dsp_matlab_Filters_initialize(JNIENV *env,
jobject this){ LR6_2_initialize(); }
```

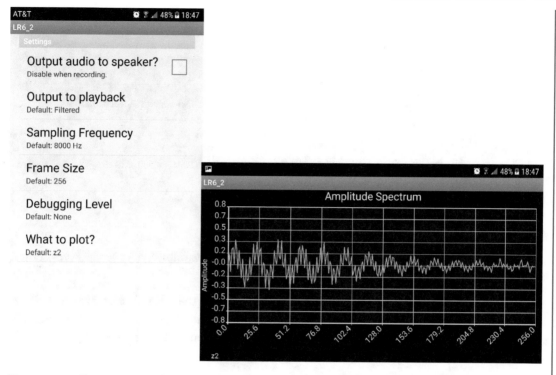

Figure 6.34: Settings view (left), output of z2, modulated input (right).

```
Void Java_com_dsp_matlab_Filters_finish(JNIENV *env, jobject this)
{ LR6_2_terminate(); }
```

One can see that all the LR5_2 references are changed to LR6_2. The compute method signature is updated to accept the correct arguments. The LR5_2 call is replaced with LR6_2 and the appropriate arguments.

If everything is done correctly you should be able see the outputs as shown in Figure 6.34.

iOS STEPS

Start a new single view project in Xcode with the same parameters used before. Copy the AudioController.* and ViewController.* from LR5_2 to the new project directory. You can also copy the file Base.lproj/Main.storyboard as well to quickly get the GUI up and running.

1. Edit the AudioController.h file in Xcode. Replace the LR4_5 references to LR6_2. From the `AudioDelegate` protocol, remove `getResistance`, `getFrequency`, `getNumCoeffs`, and `getPlotPickerIndex`. The output data length will have differ-

ent lengths so it is needed to update the `updateGraphData` method to handle this as indicated below.

```
@protocol audioUIDelegate
-(void) updateGraphData:(float*)data withDataPoints:
    (int)numDataPoints withFilteredData:(float*)filtered_data
    withFilteredPoints: (int)numPoints;
-(NSInteger) getFilePickerIndex;
-(NSInteger) getPlotPickerIndex;
-(void)didStopPlaying;
-(void)didStopRecording;
@end
```

Within the @interface code section, remove the unused variables, `numCoeffs`, resistance, capacitance, and frequency.

2. Edit AudioController.m in Xcode. First, replace all of the LR5_2 references to LR6_2, thus updating the include lines, initialize, and terminate functions. In the init method, put back the items from the audio list. Remove references to variables `_resistance`, `_capacitance`, `_numCoeffs`, and `_frequency` throughout the file.

```
_audioList = @[@"NOT_USED",
               SINE,
               SQUARE1,
               SQUARE2,
               SQUARE3,
               SAWTOOTH1,
               SAWTOOTH2,
               SAWTOOTH3,
               CHIRP1,
               CHIRP2];
```

In the `toggleMicrophone` method, override the current sample rate with the sampling rate as follows:

```
BOOL isOn = [(UISwitch*)sender isOn];
AudioStreamBasicDescription currentASBD;
currentASBD = [self.microphone audioStreamBasicDescription];
```

```
_sampleRate = currentASBD.mSampleRate;
```

Also, in the `togglePlaybackOutput` method, set the `sampleRate` to a specific value as indicated below.

```
-(void)togglePlayBackOutput:(id)sender
{
    NSINteger ind = [_delegate getFilePickerIndex];
    _sampleRate=44100.0;
    _plotIndex = [_delegateGetPlotPickerIndex];
...
    NSURL* file = [self seletedAudioFIleURL:ind];
}
```

Then, further down in the file in the method with the arguments microphone, `bufferlist`, `bufferSize`, and `numberOfChannels`, modify the calls for the output data since the output length is not the same as the input. The `appendDataFromBufferList` method will change to:

```
[self.filtRecorder appendDataFromBufferList:bufferList
withBufferSize:_frameSize];

__weak typeof (self) weakSelf = self;
dispatch_async(dispatch_get_main_queue(),  ^{
    [weakSelf.delegate updateGraphData:buffer withDataPoints:
    bufferSize  withFilteredData:_filt_data withFilteredPoints:
    _frameSize];
```

The output data size is just the frame size chosen from the GUI. This is important to remember when viewing the output data. It will be appended to an audio file as before, but the data need to be reshaped to have a matrix size `_frameSize` x N, where N is the number of frames that get processed.

Next, for the output from the audio file playback, the method with the arguments `audioPlayer`, `buffer`, `bufferSize`, `numberOfChannels`, and `audioFile` needs to be modified in a similar manner as indicated below.

```
...
[self.filtRecorder appendDataFromBufferList:filteredBufferList
withBufferSize:_frameSize];

free(filteredBufferList);
__weak typeof (self) weakSelf = self;
dispatch_async(dispatch_get_main_queue(), ^{
    [weakSelf.delegate updateGraphData:buffer[0] withDataPoints:
    bufferSize withFilteredDAta:_filt_data withFilteredPoints:
    _frameSize];
});
```

Last, the `filtererData` method needs to be modified to process blocks of data. The full method is indicated below.

```
-(void) filterData:(float*)data withDataLength:(UInt32)length
{
    int bytes_to_process=length;
    int i=0;
    int frameSize=0;
    while(bytes_to_process > 0 )
    {
        frameSize=_frameSize;

        emxArray_real32_T *x = emxCreateWrapper_real32_T(data
                        +(i*_frameSize), 1 , frameSize );
        emxArray_real32_T *con =  emxCreateWrapper_real32_T(_filt_data
                        +(i*_frameSize), 1 , frameSize+64 );

        LR6_2(x->data, x->size, _sampleRate, _plotIndex, con);

        BOOL sameptr = _filt_data+(i*_frameSize) == con->data;
        if( !sameptr )
        {
            memcpy(_filt_data+(i*_frameSize), con->data, frameSize*sizeof
            (float)); free(con->data);
        }
        bytes_to_process -= _frameSize;
```

```
    i++;
  }
}
```

3. Moving onto the ViewController.h file, update the `updateGraphData` argument signature and remove the previous get methods (`Resistance`, `Frequency`, `NumCoeffs`, `Capacitance`). Within the @interface block, update the method as noted below.

```
-(void) updateGraph:(float*)data withDataPoints:(int)numDataPoints
     withFiltered:(float*)filtered_data withFilteredPoints:
     (int)numPoints;
```

4. In ViewController.m, again remove the references to the `numCoeff`, `resistance`, `capacitance`, and `frequency` inputs and variables. Then, update the arrays that show the picker data as noted below.

```
_filePickerData = @[@"microphone",
              @"Sine 800,2000,3200 10s",
              @"Square .5Duty 100 10s",
              @"Square .5Duty 400 10s",
              @"Square .5Duty 1200 10s",
              @"Saw    .5Duty 300 10s",
              @"Saw    .5Duty 500 10s",
              @"Saw    .5Duty 1000 10s",
              @"Chirp  4000 10s",
              @"Chirp  24000 10s"
              ];

_plotPickerData = @[@"z2 mod",
              @"z3 mod+noise",
              @"y1 fft",
              @"y2 fft(mod)",
              @"y3 fft(mod+noise)",
              @"y4 fft(1/3)",
              @"z4 ifft(fft(1/3))"];
```

Update the `updateGraphData` arguments and handle the new arguments as follows:

```
-(void) updateGraph:(float*)data withDataPoints:(int)numDataPoints
        withFiltered:(float*)filtered_data withFilteredPoints:
        (int)numPoints
{
  _data = data;
  memcpy(_filtered_data, filtered_data, sizeof(float)*numPoints);
  _dataLen = numDataPoints;
  _filteredDataLen = numPoints;
...
```

5. In lab LR5_2, the sample rate and the frame size options were removed. Add those GUI elements back in the storyboard. The GUI elements need to be referenced to the code object in order for it to function properly. The second plot picker is reused and reassigned. Figure 6.35 shows a sample screenshot of the app.

6. If using MATLAB R2016b or later, it is required to add a specific header file into the *Native Code* folder separately. This header file (`tmwtypes.h`) can be found in the MATLAB root with the following path:

```
MATLAB root\R2019b\extern\include
```

LR6_3 – This real-time lab involves the noise reduction application of Fourier transform. The input signal has noise in the form of three tones added it. The user will be able control the level of the noise. Like the last lab, the user will be able to select the desired output. The available options for the output will be x2, the original signal with noise added y1, the `fft` of the signal y2, the `fft` of original signal plus noise y3, the signal with the noise reduced, and the `ifft` of y3 or noise reduced time domain signal z3.

ANDROID STEPS

These steps denote the modification of the LR6_2 lab codes. Copy the LR6_2 directory to LR6_3 and rename the files LR6_2 to LR6_3. Update the project files before updating the codes.

1. Update the applicationId in app/build.gradle to have LR6_3 in the name instead of LR6_2. Remove the old C code from app/src/main/jni and copy the new MATLAB generated code into the same directory.

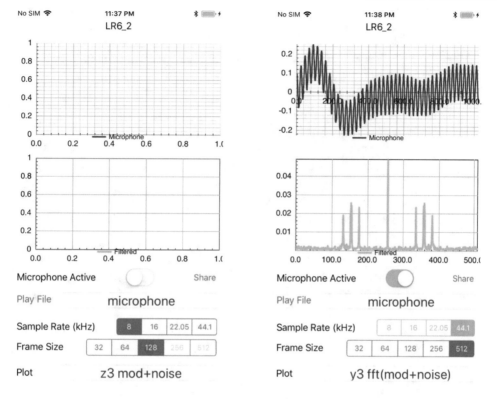

Figure 6.35: Init screen and microphone input for LR6_2.

2. If using MATLAB R2016b or later, it is required to add a specific header file into the *jni* folder separately. This header file (`tmwtypes.h`) can be found in the MATLAB root with the following path:

```
MATLAB root\R2019b\extern\include
```

3. Update the MATLAB generated files in the file CMakeLists.txt. The code section in the file appears as the following block of code. The order of the files does not matter.

```
add_library( # Sets the name of the library.
             matlabNative

             # Sets the library as a shared library.
```

```
SHARED

# Provides a relative path to your source file(s).
src/main/jni/fft.c
src/main/jni/fft.h
src/main/jni/fft1.h
src/main/jni/fft1.c
src/main/jni/ifft.c
src/main/jni/ifft.h
src/main/jni/fftshift.c
src/main/jni/fftshift.h
src/main/jni/LR6_3.c
src/main/jni/LR6_3.h
src/main/jni/LR6_3_data.c
src/main/jni/LR6_3_data.h
src/main/jni/LR6_3_emxutil.c
src/main/jni/LR6_3_emxutil.h
src/main/jni/LR6_3_initialize.c
src/main/jni/LR6_3_initialize.h
src/main/jni/LR6_3_terminate.c
src/main/jni/LR6_3_terminate.h
src/main/jni/LR6_3_types.h
src/main/jni/MatlabNative.c
src/main/jni/rtGetInf.c
src/main/jni/rtGetInf.h
src/main/jni/rtGetNaN.c
src/main/jni/rtGetNaN.h
src/main/jni/rtwtypes.h
src/main/jni/rt_nonfinite.c
src/main/jni/rt_nonfinite.h
src/main/jni/tmwtypes.h
)
```

4. Moving onto *app/src/main/AndroidManifest.xml*, update the LR6_2 names to LR6_3.

5. The file *app/src/main/res/xml/prefs.xml* will need to be edited to add the noise level option. Add the EditTextPreference as follows:

```
<EditTextPreference
      android:key="a1"
      android:defaultValue="0.1"
      andoiid:gravity="left"
      android:inputTYpe="numberDecimal"
      android:summary="Default: 0.1"
      android:title="Noise level (less than 1)" />
```

Notice that @array/choiceOptions and @array/choiceValues are still referenced, so their values need to get updated in the file *app/src/main/res/values/arrays.xml* in order to work with the new options as follows:

```
<string-array name="choiceOptions">
   <item>Noisy Signal</item>
   <item>FFT of original</item>
   <item>FFT of noisy</item>
   <item>FFT of noise reduced</item>
   <item>Noise reduced signal</item>
</string-array>
<string-array name="choiceValues">
   <item>0</item>
   <item>1</item>
   <item>2</item>
   <item>3</item>
   <item>4</item>
</string-array>
```

6. Moving onto the code changes, update the file app/src/main/java/com/dsp/matlab/Filter.java to update the function calls by adding the Settings.a argument as noted below.

```
currentFrame.setFiltered(Filters.compute(currentFrame.getFloats(),
Settings.Fs, Settings.a, Settings.choice, Settings.blockSize));
```

7. Now in app/src/main/java/com/dsp/matlab/Filters.java, update the compute method as follows:

```
public static native float[] compute( float[] in, int Fs, float a,
int choice, int frameSize );
```

8. The noise option will need to be added to the `updateSettings()` method in app/src/main/java/com/dsp/matlab/RealTime.java as follows:

```
Settings.setChoice(Float.parseFloat(preferences.getString
("a1", "0.1")));
```

9. Update app/src/main/java/com/dsp/matlab/Settings.java to store and retrieve the settings added. Nothing will be removed, only add the noise variable and set method.

```
public static float a = (float) 0.1;
public static void setChoice(float a1){ a = a1; }
```

10. Finally, update the app/src/main/jni/MatlabNative.c interface file. This file is the interface from Java to native C. The entire file is listed below.

```
#include <jni.h>
#include <stdio.h>
#include "rt_nonfinite.h"
#include "LR6_3.h"
#include "LR6_3_terminate.h"
#include "LR6_3_emxAPI.h"
#include "LR6_3_intialize.h"

jfloatArray
Java_com_dsp_matlab_Filters_compute(JNIEnv *env, jobject this,
jfloatArray, input, jint Fs, jfloat a, jint choice, jint frameSize)
{
    jfloatArray output = (*env)->NewFloatArray(env, frameSize);
    float *_output =
                (*env)->GetFloatArrayElements(env, output, NULL );
    float *_in = (*env)->GetFloatArrayElements(env, input, NULL );
```

```
//compute
emxArray_real32_T *result;
int X_size[2] = {1, frameSize};
int result_size[2];

LR6_3(_in, X_size, Fs, a, choice, _output, result_size);

(*env)->ReleaseFloatArrayElements(env, input, _in, 0);
(*env)->ReleaseFloatArrayElements(env, output, _output, 0);
return output;
}

Void Java_com_dsp_matlab_Filters_initialize(JNIENV *env,
jobject this){ LR6_3_initialize(); }

Void Java_com_dsp_matlab_Filters_finish(JNIENV *env, jobject this)
{ LR6_3_terminate(); }
```

You can see that all the LR6_2 references are changed to LR6_3. The compute method signature is updated to accept the correct arguments. The LR6_2 call is replaced with LR6_3 and the appropriate arguments.

If everything is done correctly, you should be able to see the outputs as shown in Figure 6.36.

iOS STEPS

Start a new single view project in Xcode with the same parameters used before. Copy the AudioController.* and ViewController.* from LR6_3 to the new project directory. You can also copy the file Base.lproj/Main.storyboard as well to quickly get the GUI up and running.

1. Edit the file AudioController.h in Xcode. Replace the LR6_2 references to LR6_3. From the `AudioDelegate` protocol, only add a `getNoise` method.

```
@protocol audioUIDelegate
-(void) updateGraphData:(float*)data withDataPoints:
    (int)numDataPoints withFilteredData:(float*)filtered_data
    withFilteredPoints:(int)numPoints;
-(NSInteger) getFilePickerIndex;
-(NSInteger) getPlotPickerIndex;
```

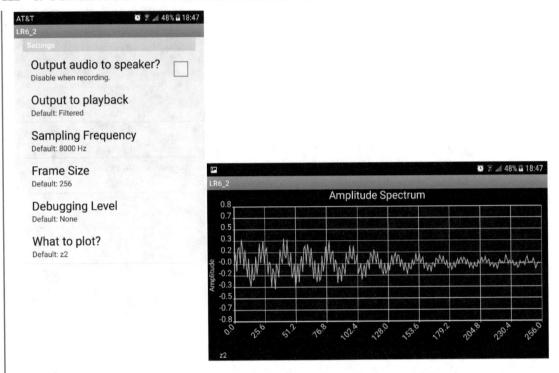

Figure 6.36: Settings view (left), output of z2, and modulated input (right).

```
-(float) getNoise;
-(void)didStopPlaying;
-(void)didStopRecording;
@end
```

In the @interface block, add a noise property to store this value as follows:

```
@property float sampleRate;
@property int frameSize;
@property NSInteger plotIndex;
@property float noise;
```

2. Edit AudioController.m in Xcode. First, replace all of the LR6_2 references to LR6_3, thus updating the include lines, initialize, and terminate functions. In the init method, initialize the noise value to −0.1 as follows:

```
...
_frameSize = 128;
_noise = 0.1;
...
```

In the method `toggleMicrophone`, add a call to the delegate to retrieve the noise value and remove the previous update of the sample rate from `AudioStreamBasicDescription` as noted below.

```
...
_plotIndex = [_delegate getPlotPickerIndex];
_noise = [_delegate getNoise];
BOOL isOn = [(UISwitch*)sender isOn];
if(!isOn)
...
```

Also, in `togglePlaybackOutput`, retrieve the noise value. Remove the hard code of the sample rate as noted below.

```
...
NSInteger ind = [_delegate getFilePickerIndex];
_plotIndex = [_delegate getPlotPickerIndex];
_noise = [_delegate getNoise];
if (self.isRecording)
...
```

The last step in this file is to update the LR6_3 MATLAB call as follows:

```
LR6_3(x->data, x->size, _sampleRate, _noise, _plotIndex, con->data,
con->size);
```

3. Moving onto the ViewController.h file, only need to add the `getNoise` method within the @interface block as follows:

```
-(float) getNoise;
```

4. In the ViewController.m, within the @interface block, add a UI textfield as follows:

```
@interface ViewController ()
...
@property (nonatomic, weak) IBOutlet UITextField *noiseInput;
@end
```

Within the init method, update the options that show in the picker view. Also, add a "Done" widget to clear the keyboard when done entering text as shown below.

```
_plotPickerData = @[@"X2 +noise",
                    @"y1 fft",
                    @"y2 fft(+noise)",
                    @"y3 fft(trunc)",
                    @"z3 ifft(fft(trunc))"];

UIBarButtonItem *barButton = [[UIBarButtonItem alloc]
    initWithBarButtonSystemItem:UIBarButtonSystemItemDone
    target:_noiseInput action:@selector(resignFirstResponder)];

UIToolbar *toolbar = [[UIToolbar alloc]
                    initWithFrame:CGRectMake(0, 0, 320, 44)];
toolbar.items = [NSArray arrayWithObject:barButton];
_noiseInput.inputAccessoryView = toolbar;
```

The last step to do here is to add the getNoise method.

```
-(float) getNoise
{
    NSString *noiseStr = _noiseInput.text;
    return [[NSDecimalNumber decimalNumberWithString:noiseStr]
        floatValue];
}
```

5. The GUI interface is the same as the previous lab except that a text input field is to be added for the user noise input. With the lack of space, you can place all the labels and

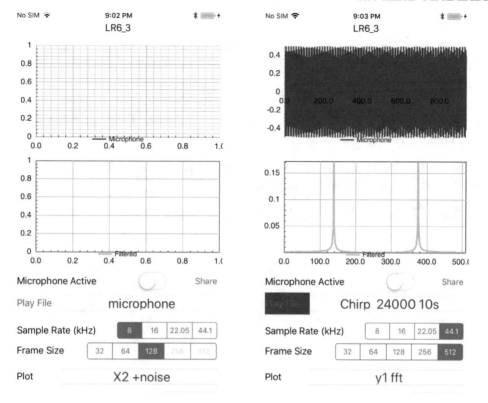

Figure 6.37: Initial screen and fft output of chirp signal.

other options in an `optionsView` within a `ScrollView`. In the ViewController.h, add objects for these elements as follows:

```
@property (weak,nonatomic ) IBOutlet UIView*      optionsView;
@property (weak,nonatomic ) IBOutlet UIScrollView* scrollView;
```

Then, in the `viewDidLoad` method, the following simple statement will allow scrolling to see all of the available options:

```
_scrollView.contentSize = _optionsView.frame.size;
```

When all of the above are done, compile and run the app. Figure 6.37 shows sample outputs.

6.6   REFERENCES

[1] J. Buck, M. Daniel, and A. Singer. *Computer Explorations in Signal and Systems Using MATLAB*, 2nd ed., Prentice Hall, 1996. DOI: 10.1109/ICASSP.2015.7178293. 177

[2] B. Lathi. *Linear Systems and Signals*, 2nd ed., Oxford University Press, 2004.

[3] D. Fannin, R. Ziemer, and W. Tranter. *Signals and Systems: Continuous and Discrete*, 4th ed., Prentice Hall, 1998. 177

[4] J. Buck, M. Daniel, and A. Singer. *Computer Explorations in Signal and Systems Using MATLAB*, 2nd ed., Prentice Hall, 1996. 198, 202

[5] O. Ersoy. *Diffraction, Fourier Optics and Imaging*, Wiley, 2006. 204

CHAPTER 7

Digital Signals and Their Transforms

7.1 DIGITAL SIGNALS

In the previous chapters, Fourier transforms for processing analog or continuous-time signals were covered. Now let us consider Fourier transforms for digital signals. Digital signals are sampled (discrete-time) and quantized version of analog signals. The conversion of analog to digital signals is normally performed by using an analog-to-digital (A/D) converter, and the conversion of digital to analog signals is normally done by using a digital-to-analog (D/A) converter. In the first part of this chapter, it is stated how to choose an appropriate sampling frequency to achieve a proper analog-to-digital conversion. In the second part of the chapter, the A/D and D/A processes are implemented.

7.1.1 SAMPLING AND ALIASING

Sampling is the process of generating discrete-time samples from an analog signal. First, it is helpful to mention the relationship between analog and digital frequencies. Consider an analog sinusoidal signal $x(t) = A\cos(\omega t + \phi)$. Sampling this signal at $t = nT_s$, with the sampling time interval of T_s, generates the discrete-time signal

$$x[n] = A\cos(\omega n T_s + \varphi) = A\cos(\theta n + \varphi), \quad n = 0, 1, 2, ..., \tag{7.1}$$

where $\theta = \omega T_s = \frac{2\pi f}{f_s}$ denotes digital frequency (with units being radians and ω with units being radians/second).

The difference between analog and digital frequencies is more evident by observing that the same discrete-time signal is obtained from different continuous-time signals if the product ωT_s remains the same. An example is shown in Figure 7.1. Likewise, different discrete-time signals are obtained from the same analog or continuous-time signal when the sampling frequency is changed. An example is shown in Figure 7.2. In other words, both the frequency of an analog signal f and the sampling frequency f_s define the digital frequency θ of the corresponding digital signal.

It helps to understand the constraints associated with the above sampling process by examining signals in the frequency domain. The Fourier transform pairs for analog and digital

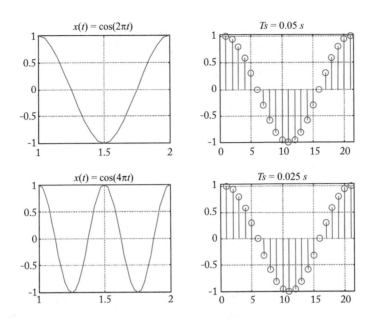

Figure 7.1: Sampling of two different analog signals leading to the same digital signal.

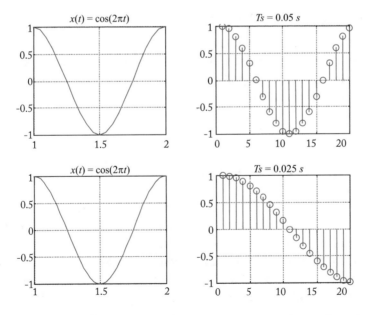

Figure 7.2: Sampling of the same analog signal leading to two different digital signals.

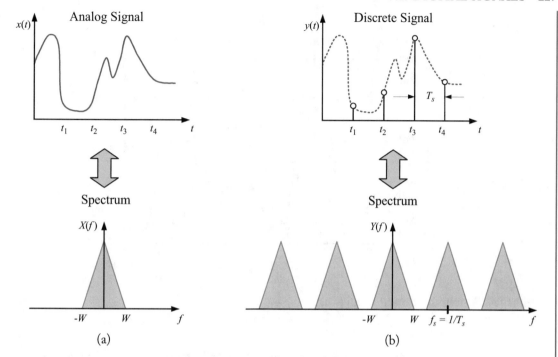

Figure 7.3: (a) Fourier transform of a continuous-time signal and (b) its discrete-time version.

signals are given by

Fourier transform pair for analog signals

$$\begin{cases} X(\omega) = \int\limits_{-\infty}^{\infty} x(t)e^{-j\omega t}\ dt \\ x(t) = \frac{1}{2\pi} \int\limits_{-\infty}^{\infty} X(\omega)e^{j\omega t}\ d\omega \end{cases} \tag{7.2}$$

Fourier transform pair for discrete signals

$$\begin{cases} X(e^{j\theta}) = \sum\limits_{n=-\infty}^{\infty} x[n]e^{-jn\theta},\ \ \theta = \omega T_s \\ x[n] = \frac{1}{2\pi} \int\limits_{-\pi}^{\pi} X(e^{j\theta})e^{jn\theta}\,d\theta. \end{cases} \tag{7.3}$$

As illustrated in Figure 7.3, when an analog signal with a maximum bandwidth of W (or a maximum frequency of $f_{\max}$) is sampled at a rate of $T_s = \frac{1}{f_s}$, its corresponding frequency response is repeated every 2π radians, or f_s. In other words, the Fourier transform in the digital domain becomes a periodic version of the Fourier transform $[0, f_s/2]$.

Therefore, to avoid any aliasing or distortion of the discrete signal frequency content and to be able to recover or reconstruct the frequency content of the original analog signal, one must have $f_s \geq 2f_{\max}$. This is known as the Nyquist rate. That is, the sampling frequency should be at

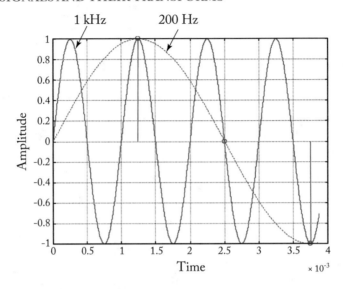

Figure 7.4: Ambiguity caused by aliasing.

least twice the highest frequency in the analog signal. Normally, before any digital manipulation, a front-end anti-aliasing lowpass analog filter is used to limit the highest frequency of the analog signal.

Let us further examine the aliasing problem by considering an under-sampled sinusoid as depicted in Figure 7.4. In this figure, a 1 kHz sinusoid is sampled at $f_s = 0.8$ kHz, which is less than the Nyquist rate of 2 kHz. The dashed-line signal is a 200 Hz sinusoid passing through the same sample points. Thus, at the sampling frequency of 0.8 kHz, the output of an A/D converter is the same if one uses the 1 kHz or 200 Hz sinusoid as the input signal. On the other hand, over-sampling a signal provides a richer description than that of the signal sampled at the Nyquist rate.

7.1.2   QUANTIZATION

An A/D converter has a finite number of bits (or resolution). As a result, continuous amplitude values get represented or approximated by discrete amplitude levels. The process of converting continuous into discrete amplitude levels is called quantization. This approximation leads to errors called quantization noise. The input/output characteristic of a 3-bit A/D converter is shown in Figure 7.5 illustrating how analog voltage values are approximated by discrete voltage levels.

Quantization interval depends on the number of quantization or resolution levels, as illustrated in Figure 7.6. Clearly, the amount of quantization noise generated by an A/D converter

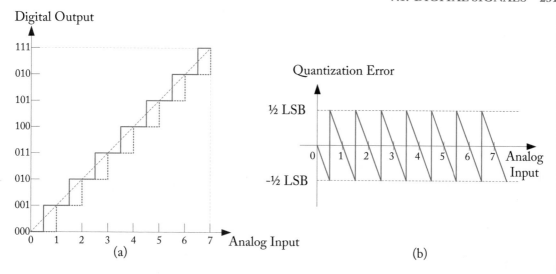

Figure 7.5: Characteristic of a 3-bit A/D converter: (a) input/output transfer function and (b) additive quantization noise.

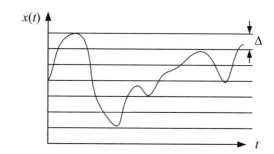

Figure 7.6: Quantization levels.

depends on the size of the quantization interval. More quantization bits translate into a narrower quantization interval and, hence, into a lower amount of quantization noise.

In Figure 7.7, the spacing Δ between two consecutive quantization levels corresponds to one least significant bit (LSB). Usually, it is assumed that quantization noise is signal-independent and is uniformly distributed over -0.5 LSB and 0.5 LSB. Figure 7.7 also shows the quantization noise of an analog signal quantized by a 3-bit A/D converter and the corresponding bit stream.

A/D Input/Output

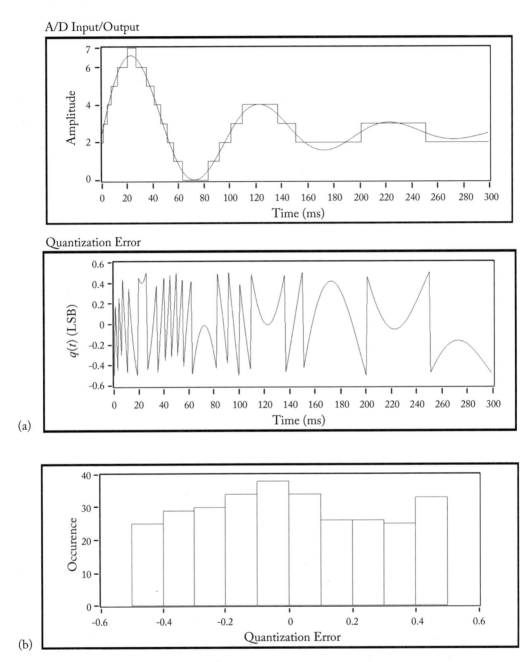

(a)

(b)

Figure 7.7: Quantization of an analog signal by a 3-bit A/D converter: (a) output signal and quantization error; (b) histogram of quantization error; and (c) bit stream.

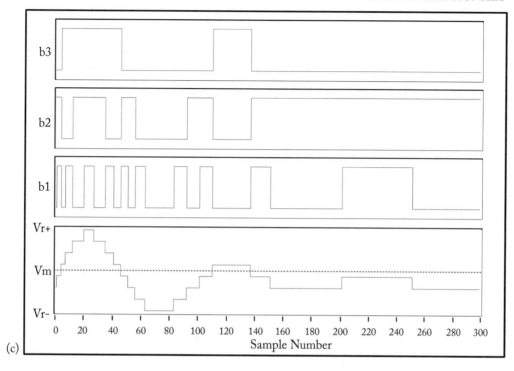

(c)

Figure 7.7: (*Continued.*) Quantization of an analog signal by a 3-bit (a) output signal and quantization error; (b) histogram of quantization error; and (c) bit stream.

7.1.3   A/D AND D/A CONVERSIONS

Because it is not possible to process an actual analog signal by a computer program, an analog sinusoidal signal is often simulated by sampling it at a very high sampling frequency. Consider the following analog sine wave:

$$x(t) = \cos(2\pi 1000t). \tag{7.4}$$

Let us sample this sine wave at 40 kHz to generate 0.125 s of $x(t)$. Note that the sampling interval $T_s = 2.5 \times 10^{-5}$ s is very short, and thus $x(t)$ appears as an analog signal. Sampling involves taking samples from an analog signal every T_s seconds. The above example generates a discrete signal $x[n]$ by taking one sample from the analog signal every T_s seconds. To get a digital signal, quantization needs to be applied to the discrete-time signal as well.

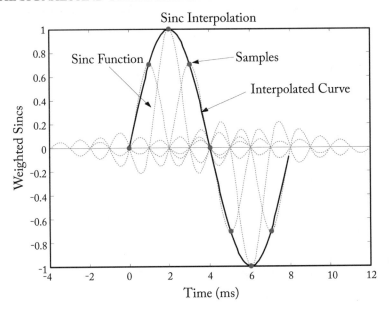

Figure 7.8: Reconstruction of an analog sine wave based on its samples, $f = 125\,\text{Hz}$ and $f_s = 1$ kHz.

According to the Nyquist theorem, an analog signal z can be reconstructed from its samples by using the following equation:

$$z(t) = \sum_{k=-\infty}^{\infty} z\,[kT_s]\,sinc\left(\frac{t - kT_s}{T_s}\right). \tag{7.5}$$

This reconstruction is based on the summations of shifted sinc ($sinx/x$) functions. Figure 7.8 illustrates the reconstruction of a sine wave from its samples achieved in this manner.

It is difficult to generate sinc functions by electronic circuitry. That is why, in practice, one uses an approximation of a sinc function. Figure 7.9 shows an approximation of a sinc function by a pulse, which is easy to realize in electronic circuitry. In fact, the well-known sample and hold circuit performs this approximation.

7.1.4   DTFT AND DFT

Fourier transformation pairs for analog and discrete signals are expressed in Equations (7.2) and (7.3). Note that the discrete-time Fourier transform (DTFT) shows a list of these transformations and their behavior in the time and frequency domains. Table 7.1 shows a list of these transformations and their behavior in the time and frequency domains.

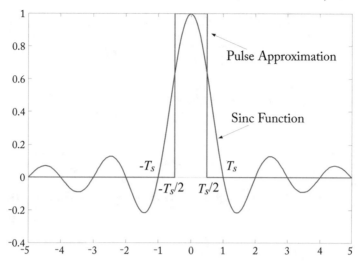

Figure 7.9: Approximation of a sinc function by a pulse.

Table 7.1: Different transformations for continuous and discrete

Time Domain	Spectrum Characteristics	Transformation Type
Continuous (periodic)	Discrete	FS
Continuous (aperiodic)	Continuous	CTFT
Discrete (periodic)	Discrete (periodic)	DFT
Discrete (aperiodic)	Continuous (periodic)	DTFT

7.2   ANALOG-TO-DIGITAL CONVERSION, DTFT AND DFT

L7.1 SAMPLING, ALIASING, QUANTIZATION, AND RECONSTRUCTION

The lab in this section addresses sampling, quantization, aliasing and signal reconstruction concepts. Figure 7.10 shows the MATLAB function named L7_1 for this lab, where the following four control parameters are defined in the code:

Amplitude (A)—controls the amplitude of an input sine wave;

Phase (b)—controls the phase of the input signal;

Frequency (f)—controls the frequency of the input signal;

Sampling frequency (fs)—controls the sampling rate of the corresponding discrete signal; and

```
Editor - C:\Users\Axa180003\Desktop\MATLAB Examples\Chapter 7\L7_1\L7_1.m        ⊙ ✕
  L7_1.m  ✕   +
1     ⊟function [xa,xd,xd1,x_recon,fn,m]=L7_1(A,b,f,fs,q)
2
3 -     theta = b*pi/180;
4 -     dt=.0001; %Sampling time for analog signal
5 -     dn=1/fs; %Sampling time for digital signal
6 -     fn=f/fs;
7 -     t=0:dt:0.2;
8 -     xa=A*sin(2*pi*f*t+theta);   %Constructing the analog signal
9 -     n=0:dn:0.2;
10 -    xd=A*sin(2*pi*f*n+theta); %Constructing the digital signal
11 -    m=dn/dt;
12
13      % Quantization
14 -    xd1=A/q*round(xd/A*q);
15
16      % Reconstruction
17 -    x_recon=interp1(1:length(xd1),xd1,1:1/m:length(xd1));
18      % x_recon=xd1;
```

Figure 7.10: L7_1 function includes sampling, aliasing, quantization and reconstruction steps.

Number of quantization levels (q)—controls the number of quantization levels of the corresponding digital signal.

To simulate an analog signal via a .m file, consider a very small value of time increment dt (e.g., $dt = 0.0001$). To create a discrete signal, sample the analog signal at a rate controlled by the sampling frequency. To simulate the analog signal, use the statement `xa=sin(2*pi*f*t)`, where t is a vector with increment $dt = 0.0001$. To simulate the discrete signal, use the statement `xd=sin(2*pi*f*n)`, where n is a vector with increment $dn = 1/f_s$. The ratio dn/dt indicates the number of samples skipped during the sampling process. Again, the ratio of analog frequency to sampling frequency is known as digital or normalized frequency. To convert the discrete signal into a digital one, quantization is performed by using the MATLAB function `round`. Let us set the number of quantization levels as a parameter.

To reconstruct the analog signal from the digital one, consider a linear interpolation technique via the MATLAB function `interp`. The samples skipped during the sampling process can be recovered after the interpolation. Finally, let us display **Original signal** and **Reconstructed signal**, **Discrete waveform**, **Digital waveform**, **Analog frequency**, **Digital frequency**, and **Number of samples skipped** in ADC as the output of the function. Use the code appearing in Figure 7.11 to verify your code and examine proper signal sampling and reconstruction by

```
Editor - C:\Users\Axa180003\Desktop\MATLAB Examples\Chapter 7\L7_1\L7_1_testbench.m        ⊙ ✕

L7_1_testbench.m  ✕  +

1 -   clc; clear; close all;
2 -   A=200; %Amplitude
3 -   b=0; %Phase
4 -   f=10; %Frequency of sinusoid
5 -   fs=2000; %Sampling frequency
6 -   q=8; %Number of quantization levels
7 -   [xa,xd,xd1,x_recon,fn,m]=L7_1(A,b,f,fs,q);
8 -   t= linspace(0,0.2,length(xa));
9     %% Verification of Results
10 -  subplot(2,2,1)
11 -  plot(t,xa);grid on;
12 -  title('Original(analog)signal')
13 -  subplot(2,2,2)
14 -  stem(xd);grid on;
15 -  title('Discrete signal (after sampling)')
16 -  subplot(2,2,3)
17 -  stem(xd1);grid on;
18 -  title('Discrete signal (after quantization)')
19 -  subplot(2,2,4)
20 -  plot(t,x_recon);grid on;
21 -  title('Reconstructed signal')
```

Figure 7.11: L7_1_testbench to verify sampling, aliasing, quantization, and reconstruction steps.

varying the parameters. First, run L7_1_testbench and examine the output signals for different parameters. Then, use the MATLAB Coder to generate the C code and incorporate it into the shell provided. (1) If using MATLAB R2016b or later, it is required to add a specific header file into the *jni* folder separately. This header file (`tmwtypes.h`) can be found in the MATLAB root with this path `MATLAB root\R2019b\extern\include` . The initial screen of the app on an Android smartphone is shown in Figure 7.12. Amplitude, Phase (degrees), Frequency, Sampling Frequency and number of the quantization level are the input parameters. The output plot can be selected from the **What to plot** picker. By pressing the PLOT button, **Digital Frequency** and **Number of skipped samples** are displayed as well as the desired plot gets shown in separate graphs.

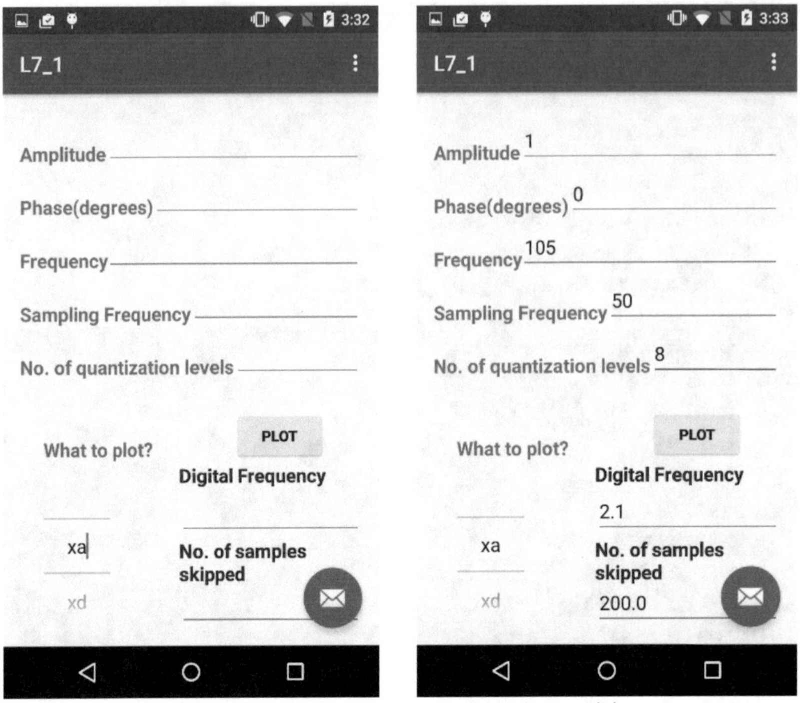

Figure 7.12: (a) Initial screen for L7_1 and (b) a desired setting for the input parameters.

L7.1.1 Analog and Digital Frequency

Digital frequency θ is related to analog frequency f via the sampling frequency, that is, $\theta = \frac{2\pi f}{f_s}$. Therefore, one can choose the sampling frequency f_s to increase the digital or normalized frequency of an analog signal by lowering the number of samples.

L7.1.2 Aliasing

Set the sampling frequency to $f_s = 100$ Hz and change the analog frequency of the signal. Observe the output for $f = 10$ Hz and $f = 210$ Hz. The analog signals appear entirely different in these two cases but the discrete signals are similar. For the second case, the sampling frequency is less than twice that of the analog signal frequency. This violates the Nyquist sampling rate leading to aliasing, which means one cannot tell from which analog signal the digital signal is created. Note the value of digital frequency is 0.1 radians for the first case and 2.1 radians for the second case. To prevent any aliasing, keep the digital frequency less than 0.5 radians.

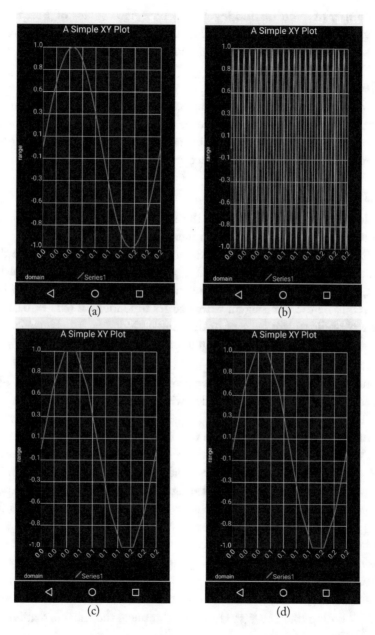

Figure 7.13: (a) Analog signals with $f = 5$ Hz, (b) $f = 105$ Hz, (c) discrete signal with $f_s = 50$ Hz generated from (a), and (d) discrete signal with $f_s = 50$ Hz generated from (b).

L7.1.3 Quantization

Next, change **Number of quantization levels** for some fixed values of **Frequency** and **Sampling Frequency**. As the number of quantization levels is increased, **Digital waveform** becomes smoother and a smaller amount of quantization error or noise is generated.

L7.1.4 Signal Reconstruction

Next, set the frequency $f = 10$ Hz and vary the sampling frequency. Observe the reconstructed waveform. Figure 7.15 shows the reconstructed signals for three different values of skipped samples. If the sampling frequency is increased, fewer samples are skipped during the analog-to-digital conversion, which makes the reconstruction process more accurate.

L7.2 DTFT AND DFT

In this section, let us compute and compare the DTFT and DFT of digital signals with the CTFT and FS of analog signals. Figure 7.16 illustrates the MATLAB function of this comparison of transforms with the verification code provided in Figure 7.17. Name the MATLAB function L7_2 and the verification code L7_2_testbench. As discussed previously, to simulate an analog signal, consider a very small time interval ($dt = 0.001$). The corresponding discrete signal is considered to be the same signal with a larger time interval ($dt1 = 0.01$).

Generate a periodic square wave with the time period $T = 0.1$. Have a switch case with **mode** `switch_expression` to make the signal periodic or aperiodic. If the signal is periodic (case 0), compute the FS of the analog signal and the DFT of the digital signal using the `fft` function over one period of the signal. For aperiodic signals, only one period of the signal is considered and the remaining portion is padded with zeros. For aperiodic signals, the transformations are CTFT (for analog signals) and DTFT (for digital signals), which are computed using the `fft` function. In fact, this function provides a computationally efficient implementation of the DFT transformation for periodic discrete-time signals. However, because simulated analog signals are actually discrete with a very small time interval, this function is also used to compute the Fourier series for continuous-time signals. Because DFT requires periodicity, one needs to treat aperiodic signals as periodic with a period $T = \infty$ to apply this function. That is why the `fft` function is also used for aperiodic signals to compute CTFT and DTFT (as done in the earlier labs). However, in practice, it should be noted that the period of the zero padded signal is not infinite but is assumed long enough to obtain a close approximation. Apply the same approach to the computation of CTFT and DTFT. Because DTFT is periodic in the frequency domain, for digital signals, repeat the frequency representation using the statement *yd=repmat(yd,1,9)* noting that the `fft` function computes the transformation for one period only.

Run L7_2_testbench and then generate the C code using the MATLAB Coder. Incorporate the generated C code in the shell provided. If using MATLAB R2016b or later, it is required to add a specific header file into the *jni* folder separately. This header file (`tmwtypes.h`) can

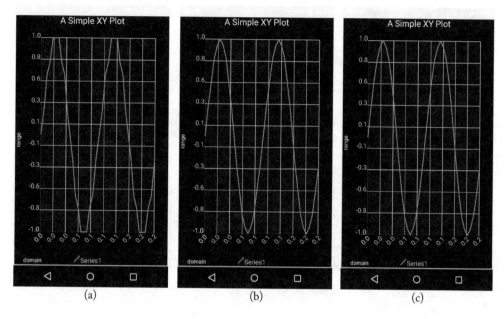

Figure 7.14: Signals corresponding to different quantization levels: (a) 8, (b) 16, and (c) 32.

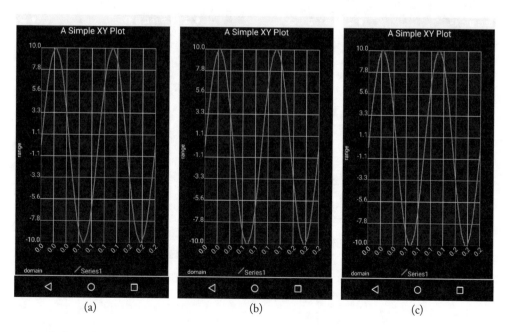

Figure 7.15: Signal reconstruction with different number of samples skipped in ADC: (a) 20, (b) 10, and (c) 5.

```matlab
Editor - C:\Users\Axa180003\Desktop\MATLAB Examples\Chapter 7\L7_2\L7_2.m        ⊙ ✕

 L7_2.m  ✕  +

1    function [xa,xd,ya,yd]=L7_2(mode)
2    T=.1;
3    f=1/T;
4    dt=0.001;
5    dt1=0.01;
6    t=0:dt:1-dt;
7    t1=0:dt1:1-dt1;
8    % xa=square(2*pi*f*t);
9    xa=repmat([ones(1,50) -1*ones(1,50)],1,10); %Analog signal
10   % xd=square(2*pi*f*t1);
11   xd=repmat([ones(1,5) -1*ones(1,5)],1,10); %Digital signal
12   ya=zeros(1,1024);
13   yd=zeros(1,1024);
14   switch mode
15   case 0 %Perodic signal
16   ya=abs(fftshift(dt*fft(xa(1:0.1*length(xa)),128))); %FFT of analog (FS)
17   yd=abs(fftshift(dt1*fft(xd(1:0.1*length(xd)),128))); %FFT of discrete (DFT)
18   type1='FS';
19   type2='DFT';
20   case 1 %Aperiodic Signal
21   xa=[xa(1:0.1*length(xa)) zeros(1,0.9*length(xa))]; % FFT of analog (CTFT)
22   xd=[xa(1:0.1*length(xd)) zeros(1,0.9*length(xd))]; %FFT of digital (DTFT)
23   ya=abs(fftshift(dt*fft(xa,1024)));
24   yd=abs(fftshift(dt1*fft(xd,1024)));
25   type1='CTFT';
26   type2='DTFT';
27   end
28   yd=repmat(yd,1,9);
29   f1=-1/(2*dt):1/(length(ya)*dt):1/(2*dt);
30   f2=-1/(2*dt1):1/(length(yd)*dt1):1/(2*dt1);
```

Figure 7.16: L7_2 function performing DTFT and DFT transformation.

```
Editor - C:\Users\Axa180003\Desktop\MATLAB Examples\Chapter 7\L7_2\L7_2_testbench.m    ⊙ ✕
 L7_2_testbench.m   ✕   +
 1 —    clc; clear;close all;
 2 —    dt=0.001;
 3 —    dt1=0.01;
 4 —    mode=1; %0: Periodic Signal, 1:Aperiodic signal
 5 —    [xa,xd,ya,yd]=L7_2(mode);
 6 —    f1=-1/(2*dt):1/(length(ya)*dt):1/(2*dt);
 7 —    f2=-1/(2*dt1):1/(length(yd)*dt1):1/(2*dt1);
 8 —    ta=linspace(0,1,length(xa));
 9 —    td=linspace(0,1,length(xd));
10      %Verification
11 —    subplot(2,2,1)
12 —    plot(ta,xa,'.');grid on;
13 —    title('Analog Signal')
14 —    ylabel('Amplitude')
15 —    subplot(2,2,2)
16 —    plot(td,xd,'.');grid on;
17 —    title('Discrete Signal')
18 —    ylabel('Amplitude')
19 —    subplot(2,2,3)
20 —    plot(f1(1:end-1),ya);grid on;
21 —    title('FS/CTFT')
22 —    ylabel('Amplitude')
23 —    subplot(2,2,4)
24 —    plot(f2(1:end-1),yd);grid on;
25 —    title('DFT/DTFT')
26 —    ylabel('Amplitude')
```

Figure 7.17: L7_2_testbench for verifying DTFT and DFT transformation.

Figure 7.18: Initial screen of L7_2 DTFT and DFT transformation.

be found in the MATLAB root with this path `MATLAB root\R2019b\extern\include`. The initial screen of the app is shown in Figure 7.18. One can select the input signal type as periodic or aperiodic and also the signal to be plotted. Figures 7.19 and 7.20 show the plots of the periodic and aperiodic signals.

L7.3 TELEPHONE SIGNAL

Now let us examine an application of DFT. In a touch-tone dialing system, the pressing of each button generates a unique set of two-tone signals, called dual-tone multi-frequency (DTMF) signals. A telephone central office processes these signals to identify the number a user presses. The tone frequency assignments for touch-tone dialing are shown in Figure 7.21.

The sound heard when a key is pressed is a signal composed of two sine waves. That is,

$$x(t) = \sin(2\pi f_1 t) + \sin(2\pi f_2 t). \tag{7.6}$$

For example, when a caller presses 1, the corresponding signal is

$$x_1(t) = \sin(2\pi 697t) + \sin(2\pi 1209t). \tag{7.7}$$

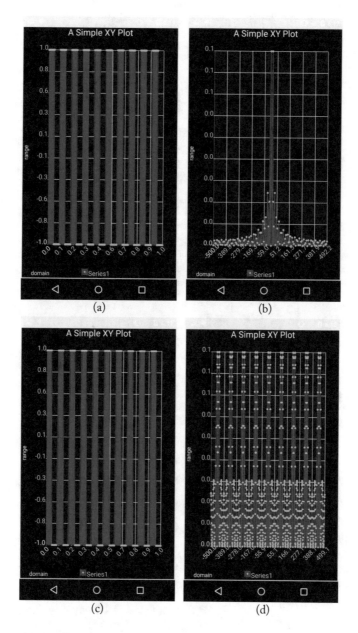

Figure 7.19: Plots of DTFT and DFT transformation for periodic signal: (a) analog signal, (b) FS transform of analog signal, (c) discrete signal, and (d) DFT of discrete signal.

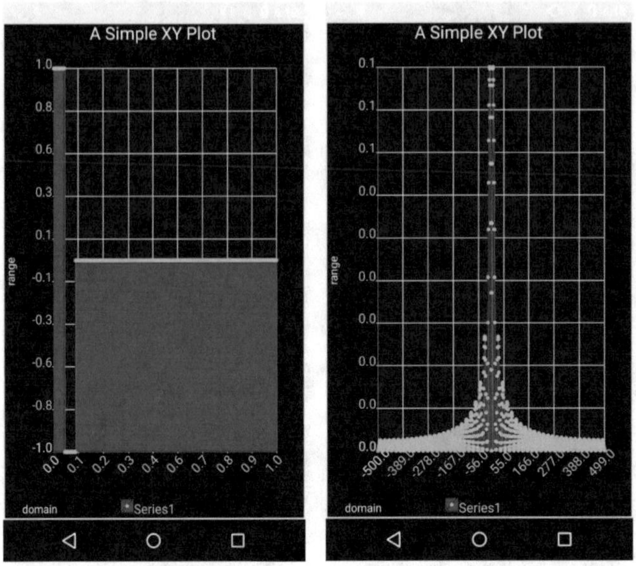

(a) (b)

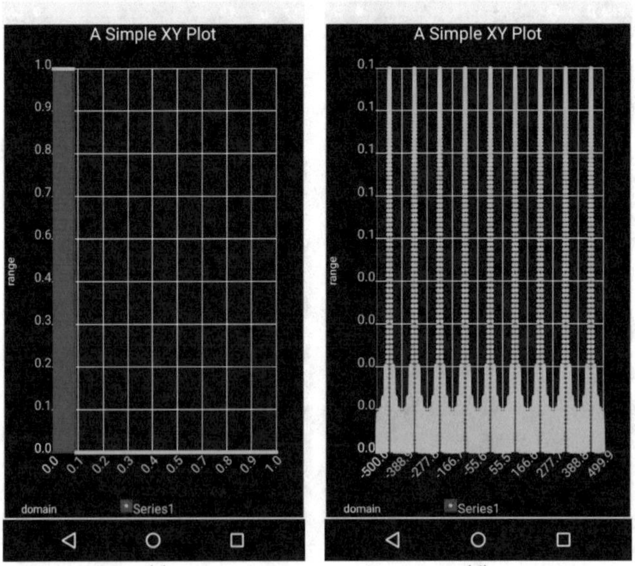

(c) (d)

Figure 7.20: Plots of DTFT and DFT transformation for aperiodic signal: (a) analog signal, (b) CTFT transform of analog signal, (c) discrete signal, and (d) DTFT of discrete signal.

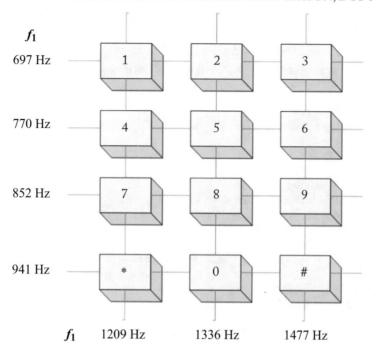

Figure 7.21: Frequency assignments for touch-tone dialing.

Table 7.2: DTMF event

Tone Type	Frequency	Timing
Dial tone	350 and 440 Hz	Continuous
Ringing tone	480 and 620 Hz	Repeating cycles of 2 s on, 4 s off
Busy tone	480 and 620 Hz	0.5 s on, 0.5 s off

Other than touch-tone signals, modern telephone systems use DTMF. Table 7.2 lists the frequency and timing for standard DTMF event signals.

In this application, let us examine the touch-tone dialing system of a digital telephone. Ten input variables $(k_0, k_1, \ldots, k_9)$ are assigned to the telephone keys $(0, 1, \ldots, 9)$. Open a new script, create a MATLAB function using the name L7_3 which has 12 inputs: $k_0, k_1, \ldots, k_{10}$ and tone. k_{10} is defined to act as a counter to count the number of times the keys are pressed; see Figure 7.22. At the beginning, when no key is pressed, the value of k_{10} is zero and the system returns the dial tone (350 and 440 Hz continuous tone). When the value of k_{10} is equal to 10, meaning that the keys were pressed for a total of 10 times, the system assumes that a valid phone number was dialed and returns the busy tone or ringing tone. Tone input parameters is

```
Editor - C:\Users\Axa180003\Desktop\MATLAB Examples\Chapter 7\L7_3\L7_3.m          ⊙ ✕

  L7_3.m  ✕  +

1     ☐ function [x,y]=L7_3(k0,k1,k2,k3,k4,k5,k6,k7,k8,k9,k10,tone)
2       % Initialize Frequency F1 and F2 based on the key pressed
3 −     if k0 F1=1336; F2=941;
4 −     elseif k1 F1=1209; F2=697;
5 −     elseif k2 F1=1336; F2=697;
6 −     elseif k3 F1=1477; F2=697;
7 −     elseif k4 F1=1209; F2=770;
8 −     elseif k5 F1=1336; F2=770;
9 −     elseif k6 F1=1477; F2=770;
10 −    elseif k7 F1=1209; F2=852;
11 −    elseif k8 F1=1336; F2=852;
12 −    elseif k9 F1=1477; F2=852;
13 −    elseif k10==0 F1=350; F2=440;
14 −    else F1=0; F2=0; end
15      % Generate the tone corresponding to the key pressed
16 −    Fs=8192;
17 −    dn=1/Fs; n=0:dn:0.125;
18 −    x=sin(2*pi*F1*n)+sin(2*pi*F2*n);
19 −    dn2=0.01; y=0;
20 −    if(k10==10) %Once 10 keys are pressed, generate ringing/busy tone
21 −        switch tone
22 −        case 0 %Ringing tone generated
23 −        n1=0:dn2:2;
24 −        y1=sin(2*pi*440*n1)+sin(2*pi*480*n1);
25 −        y2=0*y1;
26 −        y=[y1 y2 y2 y1 y2 y2 y1]; %2 sec on, 2 sec off
27 −        case 1 % Busy tone generated
28 −        n1=0:dn2:0.5
29 −        y1=sin(2*pi*480*n1)+sin(2*pi*620*n1);
30 −        y2=0*y1;
31 −        y=[y1 y2 y1 y2 y1 y2]; % 0.5 sec on, 0.5 sec off
32 −        end
33 −    end
```

Figure 7.22: L7_3 function performing touch-tone.

`switch_expression` to control the ringing tone. Keep the order of the inputs, as shown in Figure 7.22. The function has two outputs: the key pad touch tone (x) and the ringing tone/busy tone (y).

Open a new script and write a verification script for verifying the touch-tone telephone system. Name the script L7_3_testbench; see Figure 7.23. Now generate the C code using the MATLAB Coder. Incorporate the generated C code in the shell provided. If using MAT-LAB R2016b or later, it is required to add a specific header file into the *jni* folder sepa-

```
Editor - C:\Users\Axa180003\Desktop\MATLAB Examples\Chapter 7\L7_3\L7_3_testbench.m    ⊙ ✕

L7_3_testbench.m  ✕  +

 1 -   clc; clear; close all;
 2     % Values of different keys tell which key is pressed.
 3     % The one with value 1 is pressed.
 4 -   k0=1;
 5 -   k1=1;
 6 -   k2=1;
 7 -   k3=1;
 8 -   k4=1;
 9 -   k5=1;
10 -   k6=1;
11 -   k7=1;
12 -   k8=1;
13 -   k9=1;
14 -   k10=10; % Keeps the counts of keys pressed
15 -   tone=0; %0: Ringing tone, 1: Busy tone
16 -   [x,y] = L7_3(k0,k1,k2,k3,k4,k5,k6,k7,k8,k9,k10,tone);
17     %% Verifying the Results
18 -   subplot(1,2,1);
19 -   plot(x);
20 -   subplot(1,2,2);
21 -   plot(y);
```

Figure 7.23: L7_3_testbench script.

rately. This header file (`tmwtypes.h`) can be found in the MATLAB root with this path `MATLAB root\R2019b\extern\include` . Figure 7.24 shows the initial screen of the app on a smartphone.

One can select ringing tone or the busy tone to be played when the keys are pressed for a total of 10 times (a valid phone number). As soon as any number key is pressed, the corresponding key pad tone is heard and displayed. When keys are pressed 10 times (a valid phone number), the system plays the ringing tone or busy tone depending on the setting and displays the tone in the lower waveform graph.

Figure 7.25 shows the app screen for busy tone and ringing tone after pressing 10 keys for each tone type selection.

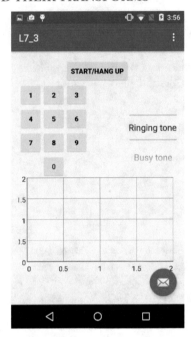

Figure 7.24: Initial screen of the touch-tone app.

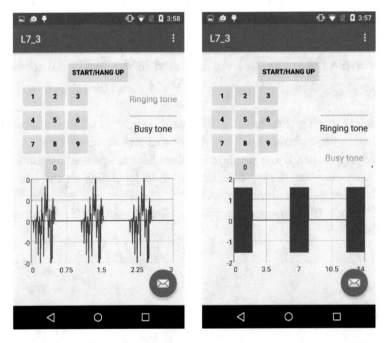

Figure 7.25: L7_3 app for a touch-tone telephone system for Busy tone and Ringing tone.

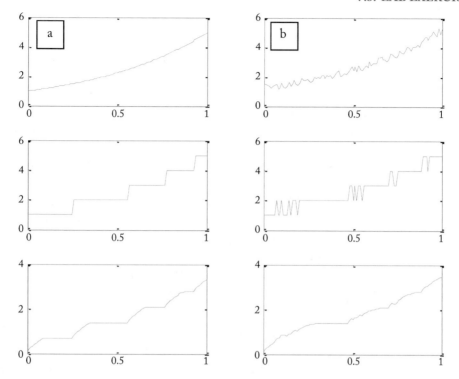

Figure 7.26: Processing at one half-step size: left column from top, the original, digitized, and smoothed signal without dithering; right column from top, the noise added, digitized, and smoothed signal with dithering.

7.3   LAB EXERCISES

7.3.1   DITHERING

Dithering is a method of decreasing the distortion of a low-frequency signal due to signal digitization [1]. Dithering works best when the sample rate is high in comparison with the rate at which the signal changes.

To see how this works, consider a slowly varying signal and its digitization, shown in Figure 7.26a. If noise is added to the original signal amplitude roughly at one half the step size, the signal will appear as shown in Figure 7.26b. If the digitized signal is passed through a resistor-capacitor circuit to smooth it out, an approximation to the original signal can be recovered. There is no theoretical limit to the accuracy possible with this method as long as the sampling rate is high enough.

Design a system to analyze the dithering technique. First, show the digitized and smoothed signal without dithering. Then, add random noise to the input signal (noise level should not exceed 50% of the step size of the digitized signal) and show the digitized and smoothed version. Measure the maximum and average error between the original signal and recovered signal.

7.3.2   IMAGE PROCESSING

DFT is widely used in image processing for edge detection. A digital image is a 2-D signal that can get stored and processed as a 2-D array. In the frequency domain, with the center denoting $(0, 0)$ frequency, the center portion of this 2-D array contains the low-frequency components of the 2-D signal or image. The edges in the image can be extracted by removing the low-frequency components.

Read and display the supplied image file *image1.jpg*. Then, complete the following steps: Compute and display the 2-D DFT of the image using the MATLAB functions `fft2` and `fftshift`. Remove the low-frequency components of the image. A user-controlled threshold can be specified to remove a varying amount of the low-frequency components. Compute and display the inverse 2-D DFT of the image using the MATLAB functions `ifft2` and `fftshift`. The processed image should reflect the edges in the original image.

7.3.3   DTMF DECODER

Design a decoder app for the DTMF system described in Section 7.3. The app should be capable of reading the touchtone signal as its input and display the corresponding decoded key number as its output.

7.4   REFERENCES

[1] S. Mitra. *Digital Signal Processing: A Computer Based Approach*, 2nd ed., McGraw-Hill, 2000. 251

Authors' Biographies

NASSER KEHTARNAVAZ

Nasser Kehtarnavaz is an Erik Jonsson Distinguished Professor in the Department of Electrical and Computer Engineering at the University of Texas at Dallas. His research areas include signal and image processing, real-time processing on embedded processors, deep learning, and machine learning. He has authored or co-authored more than 400 publications and 9 other books pertaining to signal and image processing, and regularly teaches the signals and systems laboratory course, for which this book is written. Dr. Kehtarnavaz is a Fellow of IEEE, a Fellow of SPIE, and a licensed Professional Engineer. www.utdallas.edu/~kehtar

FATEMEH SAKI

Fatemeh Saki received her Ph.D. in Electrical Engineering from the University of Texas at Dallas in 2017. She is currently a Senior R&D Engineer at Qualcomm. Her research interests include signal and image processing, pattern recognition, and machine learning. Dr. Saki has authored or co-authored 20 publications in these areas.

ADRIAN DURAN

Adrian Duran received his M.S. in Electrical Engineering from the University of Texas at Dallas in 2018. He is currently a Signal Processing Analyst at Innovative Signal Analysis, Inc. His research interests are signal and image processing, pattern recognition, and machine learning.

ARIAN AZARANG

Arian Azarang is a Ph.D. candidate in the Department of Electrical and Computer Engineering at the University of Texas at Dallas. His research interests include signal and image processing, deep learning, remote sensing, and chaos theory. He has authored or co-authored 12 publications in these areas.

Index

Printed in the United States
by Baker & Taylor Publisher Services